3D PRINTING:
A Beginner's Guide
3rd Edition

By

NIK HANDFORD

Copyright © 2019 Nik Handford

All rights reserved.

DEDICATION

To all the people who are willing to buy or make a machine that creates the parts to make it a better machine. And are willing to fail that others might succeed.

And a special dedication to their long suffering partner and especially mine, Carole, who has put up with my madness and even encouraged this one by buying me a 3D printer as a gift

CONTENTS

3D Printing:	1
DEDICATION	3
what is 3d printing?	7
Is a 3d printer for me?	11
ANATOMY of a 3d Printer	16
Preparing to print	19
What ABOUT FILAMENT?	28
Draw it up	31
What do I do with the STL file?	33
What do THE TERMS all mean?	39
Brims, Skirts, and Rafts	40
Support and Printing Orientation	43
What Now?	47
Nozzles and Layer heights	48
What next?	57
Location, location, location	64
Upgrades	65
Self Levelling Sensor	66
Noise	67
Remote Card Reader	68
Beds	69

Nozzles	70
Z motors	72
Bowden Tube	73
Cooling Fans	75
Cable Chains	76
Bed adjuster wheels/ Extruder Knob	77
Belt enhancements	78
What to look out for	82
Dangers	84
What next?	86
What tools do i need?	90
MISCELLANEOUS	92
some useful links	100
What about the future?	101
G & M codes	103
ABOUT THE AUTHOR	106

WHAT IS 3D PRINTING?

This can have a number of answers, but I will be talking about FDM or Fused Deposition Modelling. It can also be know as FFF, Fused Filament Fabrication. Or Additive Manufacturing which covers all sorts. There others but they tend to be used commercial because of the cost involved.

I think the main reason cost of FDM 3D Printer has come down is because the copyright ran out in 2010. And now even the clones of clones are being cloned, if that makes sense.

In our case, 3D printing, is basically a spool of Filament fed into a heated nozzle that is positioned over a flat bed. It's moved about and lays down a stream of material and then is raised up a set amount and it lays another stream on that until you have a 3D object. And that is it very basically. On some printer the table is dropped rather than the extruder raised but it's still basically the same machine. You will also find some printers have a single upright with an arm stretching out rather than the 2 towers with a bridge across, still basically the same machine.

There are other ways of producing 3D objects, I won't call them models because these objects can be usable devices rather than just a model of something.

When I first came across 3D printing it was about 1988, I'd spent weeks drawing up sections by hand so that a fitter could carve a wooden manifold we could test. He in turn took those section drawings and transferred them to sheets of metal to use as templates and over a number of weeks produced a wooden manifold. That was then wrapped in fibre-glass together with some metal flanges. Now he had to drill and even burn out the wood. We then tested it and got results, but we didn't really know what we'd tested after all these processes. Anyway a couple of weeks later I watched a Science and Technology TV show on the BBC called Tomorrows World and saw a stereo-lithography machine on there. Someone had created a 3d model from X-rays of a patient's damaged jaw and then printed it so that Surgeons could see where to break it and what shape they needed to create to fill the gaps.

The next day I went in to work and explained to my bosses what I saw on TV and that we needed to find away of getting our manifolds etc. produced in that way. We found a company, Formation, in Gloucestershire that could take a STL file (3D model) from CAD and produce our parts over night.

They used a tank of resin and a laser. There was a flat bed in the tank, that sat just under the surface, the laser solidified the shape for the first slice and then the table dropped a half of a mm(0.02") or so and the laser did it's thing again. This went on until your part or parts were finished. This was OK, it wasn't a great finish but with a bit of wet and dry abrasive paper it was good enough to test parts. And if they weren't strong enough we used them to create moulds from and cast metal objects. We were told British Leyland (or maybe Rover by then) were using it for their inlet manifolds. They put them on a engine to test them, until they burnt/melted and then designed the next iteration and tested that until they had the best design and then produced that one in metal.

It was great, it saved us months of work, we looked into buying a Machine but at that time it was over £750,000.

The basics of 3D printing is the same with most types of Machine, the difference seems to be the materials and system of fusing the materials together.

More recently I worked at a company that was spending a £1000 a time on manufacturing some basic microphone and remote control holders for aircraft. I suggested we buy a 3D printer as we made them from Nylon, I thought it's a no brainer as after 2 holders we'd recoup the cost of the printer and after that we'd save a fortune or rather they would make bigger profits and it was a day or two quicker to manufacture and easier to update. They rejected the idea?

Anyway we will be focused on using filament that comes on a roll, which gives us enough options;

PLA, PLA+, ABS, PLA Wood, Nylon, PC, PA, PETG, Carbon Fibre plus now metallic filaments and the list goes on and probably will continue to grow. Most of what has made 3D printing more popular is the advances in printable Materials and the ease of there use. I use 1,75mm Filament but all the same things will apply to users of 3mm Filament.

Accuracy is not the problem it once was but strength, water absorption and failure to withstand heat still can be. Everything is going to be a trade off if you want accuracy then it will be a slower print, that being said most the things you'll be able to create in hours would take a day or more any other way, so it's all relative. It takes just under an hour to print a really nice new bed adjusting wheel but I can't image how long it would take me to make it with machine tools, even if I was just doing it rough, and as for creating 4 the same I'd probably buy some. I did have to measure up what it fitted to and what it was going fit between and then I had to draw up the part and slice it, but other than the slicing which probably took a couple of minutes I would have to do the same however I made it or if I wanted someone else to make it for me. I started by using other peoples files from Thingiverse.com but found a lot were rubbish and some were great but not quite what I wanted.

The really big things are the cost and time, the adjuster wheels, that I designed and made, cost less than £0.50 which is nothing compare to getting them made, or even buying them online. I recently made some impellers, improving the design to make it pump higher, for our pond pump they cost about £0.25 to make and I could have bought a basic one online for £4.00 plus post and package. Of course I didn't have to wait they arrived 20-30mins after I sent them to the printer and I could improve the design and get another version 20-30 minutes after I'd redesigned it. And don't forget once you have that part you can print again and again, if you break the original or a friend wants one. It's not really a cost effective way to make 1000s but then it is if you don't have the money upfront to get tooling or have large batches made by others.

IS A 3D PRINTER FOR ME?

Well for me, yes, but is it for you?

People seem to have funny ideas about 3D printers, they think they can print anything. And in theory, yes, they can print most things and even things that were previously impossible, like an assembly with moving parts in it. But the misconception seems to be that you can sit down by your printer and a few minutes later you have what you want. Maybe in the future but not yet.

First what do you want to make? Has someone else already created a file(usually STL) that will print it for you? If not have you ever created an STL file and then have you ever taken an STL and created a GCODE file for a printer?

You might think you can just download the GCODE file and that's it. No. The GCODE file is the results from a Slicer program that takes the STL and creates the GCODE file purely for your Printer and Filament. Yes you could use a GCODE file that another user of the same Printer has created but it's still not guaranteed to work. You will want to create some test pieces to find the best setting for your Printer/Slicer for each material you print. I don't just mean different Materials but even for the same Material but a different Manufacturer. Even if you have the same Printer and Material as someone else they might require different settings and thus GCODE files. Honest I'm not try to put you off just saying how much goes into producing a 3D object.

Even after this you might print one object perfectly and then you do it again and it may fail. Maybe a blocked nozzle, or part comes away from the bed these things happen.

Ask yourself this;
Do I have the time/patience to first create the shape required?
Do I have the time/patience to fail printing parts before getting a good one?

If you can answer yes then read on.

First you have to decide which printer to buy, sorry no book can help you with that. Check out the latest reviews on Amazon or whatever you use and also YouTube, they have loads of "Out of the Box" reviews and "How to" videos. If you are on Facebook join some of the user groups and ask questions most will be happy to help, also you will see what users are having problems with. Be careful of letting one bad review put you off, often you'll get someone on these groups who will slag off a printer because they don't know how to set it up yet. If there are lots of the same criticism then there probably is a problem. Having said that you'll probably find someone has the answer for that so read the replies. It's unlikely that you'll find any Printer that everyone loves however good it is. More questions you'll have to answer:

1, Are you willing or able to build it from the ground up?

2, Do you want to bolt a couple of pre-assembled parts together and plug some connectors in?

3, Do you want it fully assembled and ready to plug into the mains and print.

First if you are willing and able to build it from a kit of parts you are on a to a winner. It will be cheaper and you will learn all the ins and outs of your printer and will find upgrading and fault finding a lot easier.

Second, this is the route I used with the Geeetech A30, this is not quite so cheap but you can be printing within an hour. Be warned if you take this route, make sure you check all the nuts and bolts that you didn't assemble, you will find a number of reports of users having print quality problems that turn out to be a loose joint that wasn't done up properly at the factory. Mine was no different but fore warned is fore armed as they say and I found and tightened them.

The third option, if you have money and I do mean a fair bit of money, is the fully assembled version. I don't know anyone who's gone this route but I assume it works and you should get full factory back-up if anything goes wrong as it's totally factory built. Which can be a disadvantage, had you built it you could probably find quick fixes to any problems that occur.

If you don't think you are capable of building one don't. It could be a waste of time and money, get something that is ready to use. Once you have a printer and decide you enjoy it you can always make your next printer a self-build or start modifying the one you've bought.

What size of printer do you want or need? A build area of 100x100x100, 200x200x200, 320x320x420, 500x500or1000x500 1200x600x600 or maybe 1005x1005x1005? By the time I've written this there will be bigger ones. If you're building your own you can make it the size you need. Remember the bigger the bed the longer it will take to heat although the 500x500 I've heard about apparently only has the inside 400x400 of the bed heated. Unless you think you will need a big printer it might be worth settling for something smaller and cheaper until you know 3D Printing is for you. I have a 320x320x420 but I've only used it's full Width or Length a few times and so far only half it's height probably twice. Most the parts I create are less than 25x25x50 but it has still been worth it for the few I have printed bigger parts.

Do you need or want to create object with 2 or more different colours or materials? You can get some with more than one nozzle but you can also get a device that has up to 4 rolls of material and will connect to your Printer and either pull out one Filament and feed another one in. Or cut and join the different filaments as they enter the extruder. Is this just a gimmick and something else to go wrong remember if you print with two filament even if they are the same, make, size, material say just a different colour they will probably need slightly different settings! I know users who enjoyed 3D printing in single colours and then decided they wanted a 2 or more colour printer and found it is a whole level more of difficulty, printing in multiple colours or materials. I am steering clear until enough users say it's easy. Look on forums or Facebook user groups, ask how they get on, most will give many and varied answers.

What I will say is, especially if this is your first 3d Printer, don't buy anything that has just been released, look for something that's been around for a while and getting consistently good reviews. You want something you can rely on.

"Why" you say, well when I bought my printer I'd already looked online and saw it gave good results straight out of the box. Look on YouTube and you'll find loads of people doing "out of the box" and "first print" reviews. Mine still needed modifications to function better. I'm OK with that, I expected it when deciding on it. Some bits I can print, others like stepper motor voltages can be adjusted and others like silent fans or insulated stepper motor mounts can be bought. Also when it comes to creating your print files none of the slicers(more about those later) will have a configuration for your machine so you will have to hunt for guidance on forums or wing it a bit.

The real problem comes when some Companies, and a few do, put out machines that are not fully tested. I guess we are all used to Software and Electronic Devices where the first time you use it, it says please download the latest version of software or even firmware. That's not the end of the world for most things and it cures bugs that might not been caught until thousands of user tried hundreds of different ways of using whatever it is. That's something I guess we've all come to expect. But with 3D Printers you can expect a certain amount of that on the software and firmware and even people coming out with modifications/upgrades to make the cooling better or table adjustment easier. But, without mentioning any names, some new 3D Printers won't work out of the box because they have mechanical Design faults that will be addressed later. That's no good to you, you don't want to wait until it's redesigned and you can get a new part. You want to Print Now! Even YouTube reviewers who have been given these Printers to review have refused. Some have said they could be good Printers but they had to use spares they already had from other projects to get them to work and how can you review something knowing it will not work out of the box when the general public get it!

Before buying a Printer be happy you can create the files required to make the parts you want. You will probably spend more time doing this than printing. If this doesn't appeal to you then maybe find the things you want on Thingiverse.com or somewhere and pay someone else to print them.

Creating 3D CAD/STL/GCODE files can be time consuming and you might find it frustrating and or difficult if you've never created anything like this before or you might you love it, I do.

What are the drawbacks to owning a 3D printer? Having to explain to people that no, you can't just make them a new whatever unless they can supply you with a 3d model or at the very least some drawings with dimensions.

ANATOMY OF A 3D PRINTER

Lets start at the **Filament** and go from there. So to start with you should have some sort of reel/spool holder be it a tube to go through the reel or a set of rollers the spool sit on. Whatever you have this needs to run freely and line up with your extruder.

The **Extruder** is the thing you push your filament into that drives/regulates the speed and direction of your filament. This will have **Stepper Motor** with a serrated or hobbed gear/wheel which grips the filament when it is squeeze against a sprung loaded roller. This usually has a lever on it so you release the filament to pull it out or push it in. This assembly can be remote from the **HotEnd** or attached to it. If yours is a remote Extruder then it will have a **Bowden Tube** running from it to the HotEnd. Both ends will be attached by hydraulic push fittings to keep them in place whilst the filament gets push and pulled.

The HotEnd is the place that the filament is turned into a liquid so it can be push through the **Nozzle**. The HotEnd generally comprises of 3 parts although you not be able to see any of them on most printers. The Bowden tube will enter the top of the **Heat Sink** which is most likely an aluminium body with cooling fins, this body has a thread both ends one to take the end fitting for the Bowden Tube and one to take a **Heat Break** , this joins the heat sink to the **Heater Block** and hopefully reduces the amount of heat transferred from the Heater Block to the Heat Sink. The Heater Block is again usually a block of aluminium, but can be made of other materials, and has one threaded hole going from top to bottom. So the Heat Break is screwed it to the top and the **Nozzle** is screwed in the bottom. Now to keep the heat sink cool there should be a fan, this will be on all the time. To get the heat in to the Heater Block will be a heater cartridge (probably 12-24v) which fits into a hole in the Block and to keep it under control there will be a Thermistor/thermocouple(thermostat to simplify it). On the outside of this assembly will be a **Part Cooler** often(wrongly) referred to as the nozzle cooler. Part cooler is what it does if you want modify the fan outlet to cool better make sure it hits the part as near the nozzle as possible but not the nozzle itself. The idea is to cool the filament down as it sets in place and doesn't get change to droop or sag. If you cool the Nozzle you will then have to increase the Nozzle temperature to get the filament to flow.

Chances are your HotEnd is attached to a **Carriage** which runs alone an extrusion, a rail or a pair of rods, pulled by a tooth **Drive Belt** driven by a Stepper Motor. The Drive belt will normally run from one end to the other with an **Idler Pulley** at the opposite end to that of the Stepper Motor. The Belt should be taught as it has to pull the carriage both ways and any slack will show up in your prints. This will most likely be your **X axis**.

This assembly will probably be raised and lowered by one or two Stepper Motors this time driving a **Lead Screw** this will be your **Z axis**.

Below this will be your **Bed**, it may or may not be heated, it may have three or four adjusters to allow it to be levelled and it might have none. The Bed can be made out of various materials each required a different preparation for printing ranging from nothing but a wipe over to spraying it with hair spray.

This Bed will be attach to a second carriage that moves in and out. This will most likely be your **Y axis**. As with the X axis it's most like to be a Stepper Motor at one end and an idle at the other with a toothed Drive belt.

Either built in or off to one side will be the **Control Box.**, connected to the Printer by a number of cables. Usually there will be a screen and a control knob or just a touch screen and this is where you will control it from. Some will have Wi-Fi, most if not all will printer lead socket so you can connect it direct to a Computers USB port. Also and probably more importantly there will be a **MicroSd/TF** slot so that you can plug your MicroSD card into. Setting up, Bed Levelling, and printing will all be done through the display. You could also do this through the Wi-Fi or Printer port. This contains the transformer which converts 110 or 240 volts down to 12 or 24v and a probably the most complained about part of the printer the noisy cooling fans.

This is just a snap shot of a typical printer but there are others where the Bed moves up and down and the X and Y are on rail at the top of a frame. Some are open and some will be boxed in. Some have two uprights some four and some with just one. All these machines will vary in one way or another but basically they are doing exactly the same thing, getting a line of filament down on to a surface, doing one layer then increasing the distance between the Nozzle and the Bed by a set distance and laying down another layer and so on.

PREPARING TO PRINT

Ok that said, let's assume you did do your research and saw loads of good reviews for a 3D Printer and bought one. Let's also assume you have built or assembled it. Before doing anything else go over it and check all connection are secure mechanically and electrically. Whether you did them or they were done at the factory it's easy to mean to tighten one later and forget.

Check, if the electrical connections are labelled, that they go to the correct motor.

X is the motor that drives the head left and right.
Y is the motor that drives the bed in and out.
Z is the motor or motors that raise the arm.

It's easy to get a couple round the wrong way or X is Y and vice versa. Or to think the one on the bridge is Z, it is usually the X motor. The one or two at the bottom of the towers are usually the Z motors.

Make sure the Extruder is plugged in correctly and that all the end stops Micro switches are. This is very important as most use these as 0,0,0 or home position. Plus without a working stop your motors will wreck the belt or worse as they run until the end stop Micro switches tells them to and not until it does will they stop. If you're wondering why there is only one Micro Switch for each axis it's because when you open your Slicer program it will ask you the print area size. And that will be its limiter from the end stop, hence it will normally start by hitting the end stops to find it's origin 0X,0Y,0Z or 0.0,0.0,0.0.

Before powering up, make sure the head slides left and right easily and smoothly and the table moves fore and aft equally smoothly. If they stick or are difficult, jerky then something is wrong and needs fixing before you do anything else. You can't check the vertical movement like this as it's a screw drive system which will need power so hold on.

An important thing to check before switching on, is the bridge level? The Geeetech came with 2 pieces of metal that appeared to have no function I found someone online who said they might be just to set under the bridge to level it. You might think what difference does it make the bridge will still move up and down parallel and we will set the table to be parallel to the bridge. Yes but when you take the object off the bed after printing the sides would not be square to the base. Check all cables are out of the way and do not foul on any of the moving parts.

You should have had a load of cable ties with your machine but make sure the bed and print head can move to all corners without pulling catching or kinking any of the cables.

Ok lets plug it in and switch it on. I know we haven't threaded any filament in but we are not ready for that yet.

Ok it's **ON** and has possibly moved to 0,0,0 or home but it may not have. Don't worry. Now you need to go to your menu and find "Levelling" mine has 2 options; Manual Level or Auto-Level. If you have a sensor, set it up correctly and you can use the "Auto-Level" option and it will go around to a number of positions on the bed and find its own level.

Most of us will be using the Manual method. Select this and you will see a number of positions, choose your first one and the table will move the head over that position. Make sure the read out has the thickness you are using if it isn't correct change it. Warning, changing this will change the height of the nozzle so make sure you only move it small increments, not large ones or you may plunge the nozzle into the bed. Make sure you get a piece of paper, card or feeler gauge under the nozzle and move the table using a wing-nut, thumb screw or wheel to raise or lower the bed until it drags when you push and pull on your gauge. Here is our first problem no one seems to know what this gauge is supposed to be. Some say it should be a single sheet of printing paper, other say that should be folded in two, others say use a business card! I personally would say use a feeler gauge say 0.2 mm this means you can input this in the software so it knows the true Z0 base line. Anyway when you are happy select OK and go to the next position and so on. More on this later

On mine you (which I would expect is the norm) go around the 4 corner positions and the finish with one in the middle, which you can't adjust but should be ok. Most people will advise you to now go around and do it again as moving one has a small effect on the others so it is worth going around again. Other printers may have only 3 adjusters and most small printers require no adjustment. Don't worry about the amount you have to move these, most are a M3 thread which have a pitch of 0.5 mm this means 2 complete turns of the knob raises or lowers the bed by only 1 mm.

In theory you never need to do this again but in practice you will, you'll notice something has moved when your first layer fails to stick evenly. Also if you are like me you want to make some better adjusters to either fit over the nuts or replace the nuts totally.

I created this one to slip over the existing plastic knob and it is easier to adjust it, just under the edge of the bed instead of groping around under the bed, it's also better for fine adjustments than the old knob. You can push it straight on the knob or take the knob off and push it through, then when you thread it back on it can't ever fall off. Don't forget, when you do this you must re-level the bed. That means check all positions again not just the one you changed at that time.

As an update to this I now enter 0.1 mm on my levelling screen and use a 0.15 mm feeler gauge to adjust the corners. In my slicer settings I use Z offset as 0.0 and a first layer flowrate of 115%. This seems to work well. I mostly use 0.2 mm as my first layer and always use the brim command and watch whilst it's printing. Obviously if your nozzle is too low then little to no filament will come out, if the brim comes out as individual strands then the nozzle is too high stop it and adjust your heights and start again. The strands should become one piece as the layer is printed. After printing I check the thickness of the brim to see how close I am to 0.2 mm and it's usually plus or minus 0.03, which I'm ok with. If it is consistently bigger or smaller than this either use a different size feeler gauge to compensate or a different Z value in your levelling display. So if you set 0.1 mm on the Display and use a 0.15 mm feeler gauge and it measures 0.3 mm when you check the brim. Use a 0.05 feeler gauge and check again.

As I said in theory you won't need to level the bed again but in practice you will. If you have a heated bed you may want to check it at the temp you are going to be using. I didn't when I tried PETG which require more bed heat than I'd used before. It wouldn't stick to the bed then I remember someone saying check the level at the temp your going to be printing at and sure enough the whole bed was too low. I've found if my first layer doesn't stick I check the bed temp and then the Nozzle height.

Speaking of bed temperatures I guess I'd better mention bed preparation. On my A30 it has a heated and non stick when cold glass bed. The object sticks and then when the bed starts to cool it comes away in your hand, no need for a scrapper. Not all beds are going to be like this. Some you have to prep, it could be spray with hair spray(I kid you not), smear with a glue stick or PVA/water mix, cover in blue painters tape and there are more. You'll need to find the one that works on your Printer. And remember to clean the bed regularly if there are any greasy hand prints or debris from previous prints, on the bed or tape, PVA, or whatever you're printing on, your Prints won't stick. I use alcohol wipes and every now and then acetate. I've read and heard you don't need acetate as it won't have any effect on PLA but it cleans my bed.

One more thing to check before we think about printing. Now the power is on we can go into Control and then Move function and move the bridge up (+z) and down(-z), using the controller, a few times to check it moves freely and smoothly. And the same with right (+x)and left(-x) and don't forget out(+y) and in(-y). If any of these are jerky then they need looking into before continuing.

I guess you want to print something now. As with most, mine came with 2 demo files on it's MicroSD card, these told me nothing about what they were, mine were labelled A30 and a number-.gcode so I didn't know the size or what they were.

You may find nothing shows up on your Printer, many users have found there are demo files on the MicroSD/SD card but in a directory/folder. The Printer will only see what is in the root Directory/Folder of the card. So check it on your Computer and move any you find into the cards root Directory/Folder.

Back to the internet I went on to

http://gcode.ws/

dragged the first file in there and it told me

Model size is: 89.49x46.01x71.80mm
Total filament used: 12865.72mm
Total filament weight used: 32.18grams
Estimated print time: 3:0:44
Estimated layer height: 0.20mm
Layer count: 360 printed, 363 visited
Time cost: 3.01
Filament cost: 1.61

And it showed me a 3D image of Benchy the Bench Mark Boat.

The second appear to be a Buddha

Model size is: 77.85x69.33x174.20mm
Total filament used: 42581.69mm
Total filament weight used: 106.50grams
Estimated print time: 8:57:50
Estimated layer height: 0.20mm
Layer count: 872 printed, 875 visited
Time cost: 8.96
Filament cost: 5.32

Given that the Machine came only with a few loops of filament I choose the smallest, Benchy, even though it still needed 12.8 metres of Filament and I didn't have any where near that much.

Now it's the time when you open the filament and thread it through. Hopefully you haven't already opened your filament especially if it's a 1kg reel, "Why?" you ask. It hydroscopic, it loves and craves water, you need to keep it sealed and with something that loves water even more like silica. If you don't you may find in bad cases you get pops and hisses as you are printing and your object may have small voids. You can fix this by leaving the reel of Filament in an oven at low temp for up to a day! Its mostly OK when printed but keep it dry in Filament form. I've used PLA for Pond Pump Impellers with no problems so far.

Check the end of the filament, if it has a squared off end cut it to give a bit of a taper. This helps it find the hole to the extruder as will making sure the first 50mm (2 inches) or so is fairly straight. If it's too curved it will miss the hole the other side of the feeder wheel. Alright feed the Filament though, mine had a Filament detector to stop the printer if it runs out, but only if you go into the setting and turn it on!(They don't tell you that) Next it goes into the feeder and then on to the extruder, there should be a lever to move the feeder drive wheel out of the way. Now you push the Filament in to the hole until you feed in enough to reach the extruder and release the drive wheel. If it stops just the other side of the drive wheel wiggle it about in your fingers until it goes in the hole. If it won't, withdraw it and check that it is straight as you can get it and that it has a good clean end cut at an angle, and try again. Once it get to the other side press the lever again and feed in the filament as far as you can. If it's direct drive it will only go in about an inch and then stop. With a remote drive you'll need to feed in enough to go around the Bowden tube and reach the hot end.

Right this is it, Power is on, bed was levelled, Filament is in, go to your control panel and this time you want to heat up your nozzle to say 200ºC and once it's up to temp push the filament through until it comes out of the nozzle(remember to release the extruder arm whilst doing this). Either with a pair of tweezers, long nose pliers or small scissors break/cut off the filament from the end of the nozzle.

Back at the control panel select Print and choose the model and select Print. Wow it worked, it ran out of Filament half way through but the quality was good.

Well that's what I'd like to have happened but it was not quite like that. Having realised I didn't have enough Filament to complete either of the test pieces I went on to

https://www.thingiverse.com/

and downloaded a small file for a bag clip so what do you do now you can't print an STL file. I had to convert it to GCODE, so I downloaded Cura

https://ultimaker.com/en/products/ultimaker-cura-software

Before you even download it, it first asked what Printer you are using but being a newish printer it didn't have it in its list. So I downloaded Slic3r from

http://slic3r.org/

ran it and had to guess what values were, then dragged my STL file into that and exported a .gcode file. I copied that on to the MicroSD card and then plugged it into the printer. Finally we are ready to print. It started off OK and then I realised that the coil of Filament I was given with the Printer was not in fact one piece but 2. Of course I didn't spot this until I noticed there was nothing coming out of the nozzle. I hit the pause button and spotted the problem, I pulled the piece of Filament back out of the extruder and put the end of the second piece in. I hit the pause and after warming up again it was off and running, although now printing 0.2 mm or so above the last layer, oh well it might as well finish now.

Right what had I got, amazingly a working bag clip. It wasn't brilliant a couple layers of outside bottom strands were just that but the main part was goodish if you ignore the void due to the lack of anything 2/3rd through. After looking online I concluded the bed temperature might be a bit low and or the nozzle set too far above the bed initial when I did the level, hence I think it would have been better to use a feeler gauge and put that dimension in to the setting.

All in all it wasn't bad for an first attempt.

Now I re-levelled the bed using single thickness of paper to make sure the next first layer would stick. Then I loaded the A30 Benchy test model and pressed "PRINT". Because this was set up for the A30 it had different temperatures to the ones I had in my first test. As I said before it came out pretty good. By no means perfect but it didn't have steps in it, it wasn't blobby. I could stand on it with 14 stone of weight and it didn't fall apart or crush which surprised me, so it had fully bonded together.

WHAT ABOUT FILAMENT?

Well the first thing, you'll want to get some filament and start printing stuff, so first you have to decide what Filament to buy. The obvious one is PLA I certainly assumed the bits I got with mine were PLA and so it would seem the thing to continue with but it also depends what you want to do with that print.

So below is a list of just some of the Filaments available and some of their attributes.

PVA water soluble seems mainly to be used in twin filament printers so that you can print soluble support material that will wash off after printing if you use this you need a bed temp of 45-60ºC and Nozzle temp of 185-200 ºC.

Polycarbonate good on strength, and about best in durability, above average in stiffness if you use this you need a bed temp of 80-120ºC and Nozzle temp of 260-310ºC. These temperatures may well put it out of the range of most printers, not to worry see upgrades.

PETG is about average in strength, and quite durable, and average again in stiffness if you use this you need a bed temp of 70-90ºC and Nozzle temp of 230-250ºC. PETG and Polypropylene are the only Filaments that are supposed to be Water Resistant

Wood Filled is again about average in strength, but not very durable, but it is good in stiffness if you use this you need a bed temp of 45-60ºC and Nozzle temp of 190-220ºC

PLA is above average in strength, and stiffness but below average in durability, if you use this you need a bed temp of 45-60ºC and Nozzle temp of 190-220ºC

ABS is just below average strength, and quite durable, and average in stiffness if you use this you need a bed temp of 95-110ºC and Nozzle temp of 220-250ºC

Metal Filled is below average in strength, and durability, and about the stiffest thing around if you use this you need a bed temp of 45-60ºC and Nozzle temp of 190-220ºC

ASA is just above average strength, and very durable, and average in stiffness if you use this you need a bed temp of 90-110ºC and Nozzle temp of 235-255ºC. This is the only Filament which is said to be UV resistant but lots of people have left various materials out and after a year some become more brittle and most fade but nothing disastrous happens.

There are loads of others, Nylons, Carbon Fibre Filled, HIPS, lots of PLA blends(to give higher strength or heat tolerance) and flexible ones, they all have there own pros and cons obviously if you just want to proof an object you can do it in ABS or PLA as they are the cheapest. If you use a different material to proof the object don't use the same Gcode file or your temps or speeds might be wrong.

Don't worry too much about the properties if you look on the web lots of people have created things in ABS or PLA and say that they have been using them for years and they haven't failed. Don't forget if they fail after a few years you can always use the same file to print them again.

You may think I want to use Polycarbonate but my printer HotEnd only gets up to 250ºC and the bed only goes up to 100 ºC. As you may have guess this is not a problem there are so many bits to upgrade your Printer the sky is the limit. Almost any part of your Printer can either be redesigned by you, downloaded and printed by you or bought over the internet. More of this later.

The thing most likely to damage your prints is heat, i.e. you make a phone holder for your car and leave it there on a sunny day. If you printed it in PLA it might deform but people have discovered you can increase it's resistance to heat by putting it in an oven for a time at 80-100ºC. You may find that the object is now shorter and wider so if you find the need to do this you may want increase the height(Z), and decrease the width(X) and length(Y) by a couple of percent.

Don't worry if you haven't got a heated bed as I said before there are things you can put on your bed to make it stick.

Don't blindly follow the manufactures temperatures or anyone else's, these are a guide or starting point and nothing more. They may be correct but your Thermistor may be off by 10 or more degrees. So always do some test pieces when you use a new roll of filament, even if it's the same as one you had before, they can vary.

DRAW IT UP

Ok so you decided on the Filament how do you get the file for the object you want to print?

First port of call would be the internet, has someone else wanted the object and created an STL(stereolythography) file? Go to

https://www.thingiverse.com

and search there. Chances are you won't always find exactly what you want there, but it's full of ideas you can adapt. You can also find models at

https://3dwarehouse.sketchup.com

To create your own models or modify existing ones you'll need a 3D CAD (Computer Aided Design) package or something that will allow you to build solid models and output an STL file.

There are lots of free packages and even online CAD type packages that will allow you to create your objects. The obvious ones that spring to mind are Tinker CAD which is a free online CAD package. Fusion 360 which is free if you are not using it for commercial work. Or Blender 2.8, this is not a CAD package as such it's more for artistic modelling. Which ever App/Program you use, once you are happy with it, you need to do a Save As and select .STL file.

A word of warning, I have spent years producing digital/electronic or 3D Computer Models and the one mistake lots of people make is to create their CAD model to some scale or another. Don't. Always create Drawings and CAD models 1 to 1 (full/life size). You can scale it when you print if needs be, but once you do anything to a scale it's easy to get confused.

And another thing when you create a part, measure it. Adjust the Slicer if it comes out the wrong size, not the CAD model or else your .STL file will only work for that printer and if you ever correct it later, all your earlier work, should you need to print them again will need changing to get the correct sizes.

So create everything 1 to 1, especially if whatever it is attaches to something else. Once you are happy with your Design save it as whatever your CAD program wants to and then do "SAVE AS" and save it as a .STL file.

If you're making a number of parts to create an assembly try to create them in an assembly and then save them as individual parts so that you know they fit together, rather than Print them all out and then find they don't work together.

Know your Printer if you use a 0.5 mm Nozzle then make sure wall thicknesses are a multiple of that, this is not important on models with thick walls with infill but any solid walls need to be a multiple of your Nozzle diameter or else you will get some voids in them.

The same goes for any horizontal features make sure they start and end on a level you are printing at. i.e. if you have a first layer of 0.2 mm and the rest at 0.28 then features need to be at 0.2, 0.48, 0.76, 1.04, and so on.

"Advertisement", on Amazon you can find the "3D Printer Calculator" App by Omni-Auto's which will calculate, Optimum Z step, E step corrections, X - Y corrections, Primary Filament Flowrate, Optimum Wall Thicknesses, Optimum Layer Thickness and all layer heights.

WHAT DO I DO WITH THE STL FILE?

You now need a Slicer. You may have thought you'd created your model so that's it, send it to the printer, job done. Sorry that .STL file contains

facet normal 9.990483e-001 -4.361778e-002 0.000000e+000
outer loop
vertex 3.954287e+001 2.058141e+001 3.000000e+000
vertex 3.954287e+001 2.058141e+001 1.000000e+000
vertex 3.958092e+001 2.145296e+001 3.000000e+000
endloop
endfacet
facet normal 9.990483e-001 -4.361778e-002 0.000000e+000
outer loop
vertex 3.958092e+001 2.145296e+001 3.000000e+000
vertex 3.954287e+001 2.058141e+001 1.000000e+000
vertex 3.958092e+001 2.145296e+001 1.000000e+000
endloop
endfacet
facet normal 9.914449e-001 -1.305255e-001 0.000000e+000
outer loop
vertex 3.958092e+001 2.145296e+001 3.000000e+000
vertex 3.958092e+001 2.145296e+001 1.000000e+000
vertex 3.969479e+001 2.231789e+001 3.000000e+000
endloop
endfacet
facet normal 9.914449e-001 -1.305255e-001 0.000000e+000

outer loop
vertex 3.969479e+001 2.231789e+001 3.000000e+000
vertex 3.958092e+001 2.145296e+001 1.000000e+000
vertex 3.969479e+001 2.231789e+001 1.000000e+000
endloop.

And so on for the next 40000 lines and that's for something simple.

Your Printer can't use that it needs Gcode instructions like this

M221 T0 S90.00
M140 S60.00
M104 T0 S205.00
M109 T0 S205.00
T0
M190 S60.00
G21
G90
M82
M107
G28 X0 Y0
G28 Z0
G1 Z15.0 F6000.0
G92 E0
G1 F140 E29
G1 X20 Y0 F140 E30
G92 E0
G1 F6000.0
M117 Printing...

Don't worry you don't have to know any of these Gcode commands, although it can be useful, the slicer will do all of this for you. If you're interested some common G and M codes can be found at the end of the book.

So what is a Slicer?

A Slicer is a program you can import/load your STL file into, then if you tell it about your Printer, Filament and things you require, it will literally slice it into layers the Printer can print and it will contain all the instructions to set temps and feeds and speeds of printing. Then normally after showing you a preview it will export the Gcode file that you can send to your printer or put on a MicroSD card and load into your printer to print.

Most printer documents will point you to a Slicer to download, mine was **EasyPrint3d** for the Geeetech which seems to be **Cura**, but I tried **Slic3r** and finally used **ideaMaker**. They all seem to be much of a muchness but ideaMaker seemed to work best for me on my PC.

Whichever you use the first thing you have to do is choose your printer. If yours has been around a while this will be easier as it's probably already in the list, mine wasn't so I had to create a new profile for it. This is fairly basic Name, Build Dimensions X Y and Z, max temp for nozzle, it's nozzle size, does it have a heated bed and if yes what's it's max temp.

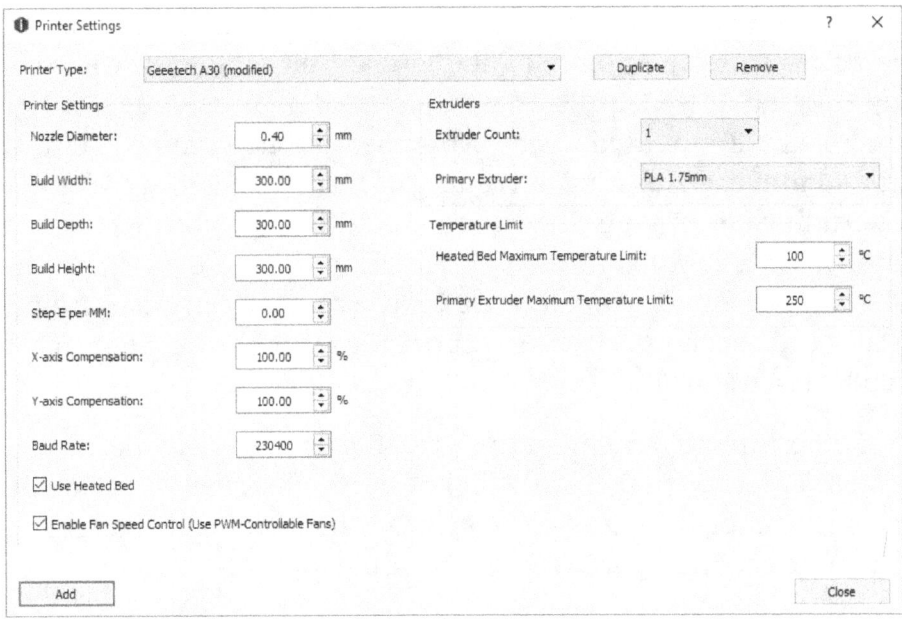

Again we have to thank the internet because that's where I got most of my initial settings. Even if it's not listed someone will have probably put the settings on the internet somewhere and if not most of the newer printers are based on one that went before so you can try those settings.

OK it will now want to know what Filament you are using and it's thickness. Check the diameter in a few place with a Vernier or Micrometer if possible as just because it says 1,75 it might not be 1,75. It will also probably ask what temperature you want use for the nozzle and bed. These will only be starting points, if you print some of the test pieces this will show you if you need to reduce or increase these temperature, but this is only for this Filament from this Manufacturer. What works for one Manufacturer may not work for another. What works for one type of Filament is unlikely to work for others, check on each reel they give you a range of temps for the bed and nozzle. Even these may not work for you as there is no guarantee that the temperature your printer says the nozzle and bed are is true. This why your Gcode may only work for you on one Printer with one Filament.

This is where you can correct any size problems or feed problems, measure off 150mm of filament and either with the Nozzle up to temp or Nozzle/Bowden disconnect go in to your settings and feed 100mm of filament. Now measure how much is left, if it's 45 mm then it's over extruding or if there's 55 mm it's under extruding. So in your Slicer you would change your Extrusion Multiplier(or Primary Filament Flowrate) from 1 to .95 if it was feeding in 5mm too much or 1.05 if it was 5mm too little. It will still need checking with actual printing results to best correct it this is just a starter.

Something to remember for later. If this doesn't work for you create a part with 0.4 wall thickness, or whatever your Nozzle size is, and print it check the printed thickness and change the flowrate accordingly. So if the wall comes out .42.

0.42 / 0.40 = 0.952

If your Extrusion Multiplier started out as 1 then it will now be

0.95, if it was already .96 then multiply .96 by .95 and enter 0.91.

Another way some users have said they do it is to change the filament diameter.

1,75 x 0.952 = 1,66

Put in Filament diameter 1,66 and Slice it again and check the thickness.

You may think it sounds wrong to have figures like this but when I changed my Nozzle to a 0.6 mm, I had to use 0.75(75%) as the Primary Filament Flowrate for PLA and 80% for PETG.

An Update to this I've, been buying the same make of PLA filament just different colours and no two are the same flowrate. The lurid green I just got is down to a Filament Flowrate of 55% to print correctly. I use the single wall thickness check and that seems to work well.

You may be thinking why don't you connect your Computer to the Printer like you do with a normal printer. Yes you can and the Slicer will send the Gcode commands straight to it and print. I don't know anyone who does. From my own point of view I want to eliminate all the things that can stand between me and a successful print. So if I put the Gcode on a MicroSD card and put that in the printer I only have to worry about that. If I have the Computer talking to the Printer I've introduced a whole other level of possible faults that might lead to failure. Plus do you want to tie your Computer up for hours or even days when you don't want to use it in case you upset the printing. I'm sure it would work fine, and by all mean connect your Computer to your 3D Printer, that's your choice. Also after the initial amazement of watching your

NIK HANDFORD

design immerge from the Bed dies down a bit you might want to work on your Computer in peace and quite without the possible smell of some of the materials you might want to use.

WHAT DO THE TERMS ALL MEAN?

Layer Height, this will be the thickness of the printed material, so if you print with a layer height of 0.3 mm it will print quicker than if you print at 0.2 or 0.1 but the quality and detail won't be so great. Generally if you're creating a one off then you do it the best quality but that might take a while and especially if your not sure it will do the job you might want to do a course/draft print and if it's what you wanted then reduce the layer height and increase the quality. It's just the same as draft and document print quality on a normal paper printer.

First Layer Height, why is this different? Well you want to make sure it has a good bond to the bed so you make it a thinner layer so it will fill any deviation in the bed and give a good layer to print on. Generally make your first layer 80-90% of your normal layer.

BRIMS, SKIRTS, AND RAFTS

You'll see these in the Slicer and you may or my not want them but this is what they do.

Lets start with a raft, the Printer will print the area under you print with a few layers to build a raft for your print to sit on. This will set a level base and hopefully stick well to the bed. It also increases the base area of your print, if it's too small to stick by itself. You can tear this off after printing.

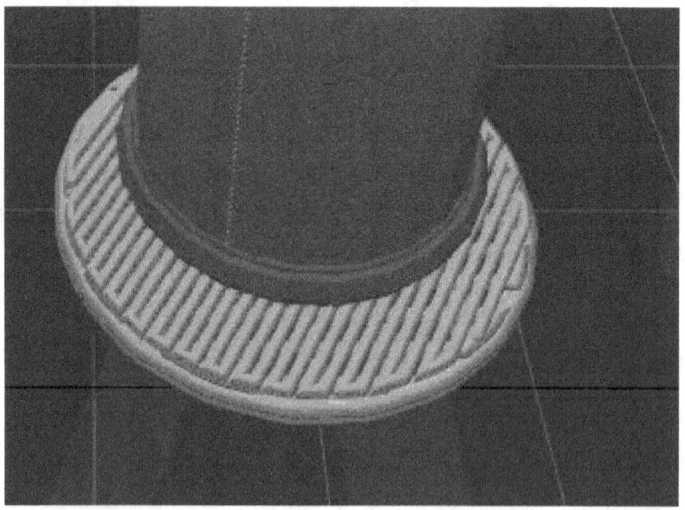

Skirt, this is a number of boundary lines around the object. You can say how many lines and how far away from your object it is. Why you may ask, this will allow you to see that the filament is running well, and the bed is level and at the correct height before it starts on the base of your object.

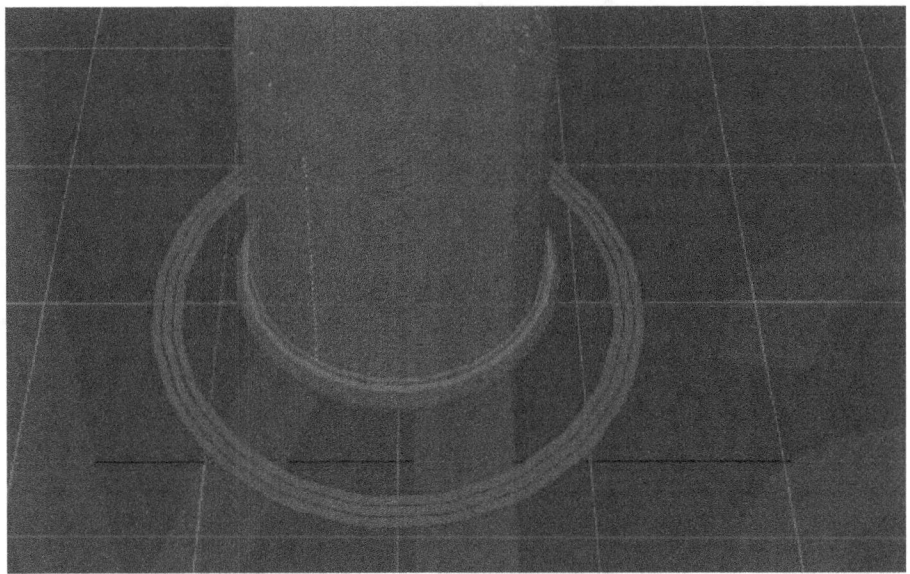

Brim this is similar to the Skirt but is connected to the base of the model and again can be ripped off after removing from the bed.

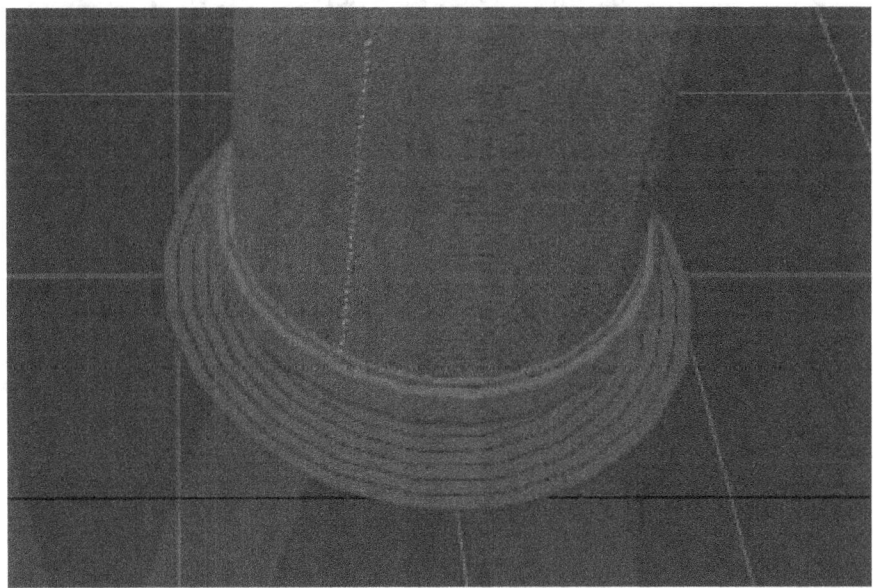

Why would you do this? As before it will start printing here so you can see a good flow before it starts on your object. Also this will increase the base area so might make an object with a small base area printable.

A couple of useful things with brims or skirts is to first watch them as the print, if they come out as individual strands the nozzle is too high. If they go transparent and/or vanish in places the nozzle is too low. Hopefully they look the same and merge into each other. And second is to measure the thickness (height) of the brim or skirt after printing, it should measure the dimension you put in your Slicer as your first layer thickness. If it doesn't alter the paper thickness you use to level the bed. Thicker if it come out too thin and thinner if it comes out to thick. This is why I level my bed using feeler gauges (available for car maintenance)

If your first layer was supposed to be 0.2 mm and it comes out 0.14 mm and you used a 0.1 mm feeler to get that then use a 0.15 mm and it should be as near as you'll get it.

If you're using a piece of card or paper this is obviously harder to do accurately so the other way is to change the figure on your display. i.e. use the same piece but when your levelling, and if the height is set at 0.1 mm reset it to 0.05 mm and go round and readjust the bed height.

You can also use Raft with Skirt or Raft with Brim, or None.

SUPPORT AND PRINTING ORIENTATION

Not the cheerleader kind, more the kind if you have something that requires part of it being printed in thin air or overhanging over 45º it will need some support or and this is worth thinking about can you rotate your object in your Slicer so you don't have any unsupported structure or overhangs over 45º. Here is an example as seen in the 3d modeller to the left and Slicer on the right.

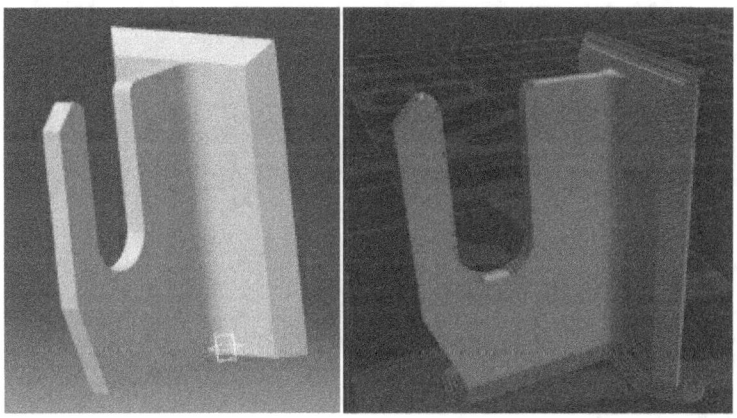

It is a net curtain rod hanger, the obvious way to do it is to sit it flat end on the bed, but it may not be the quickest or best way to do it but it would leave us with no overhangs or anything unsupported. This the way I drew it to print it in the vertical.

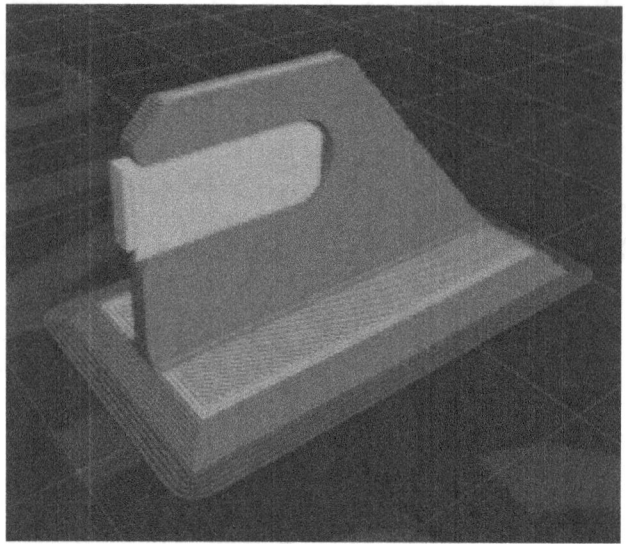

Probably for the best looking result and quickest print I need to rotate it and have the flat base on the bed, this also means I don't have to have a flat end. As you can see this will require support or the opening will collapse.

Because I've cut the end at a 45º angle I also get a 3rd option, to print with the angle edge flat to the bed.

This isn't a great example as I have 45º chamfers going to nothing around the edges but then perhaps that makes it a good example. If you look at the Slicer images you may notice even the Slicer is having problems with the sharp edges in 2 of the set-ups . If you look at the vertical and 45º ones the sharp edges look uneven and sure enough when I printed them out they looked no where near as good as the flat one. Not helped by the fact I think I'm over extruding the filament I should have done a test first but though I'd do the 3 on the same setting whatever they turned out like.

Having said that the flat one came out pretty good, the vertical one is stringy with a velvet look about it and the 45º just looks rough.

I have to say when I started I'd thought the flat one with support wouldn't be worth the extra filament but I have to change my mind

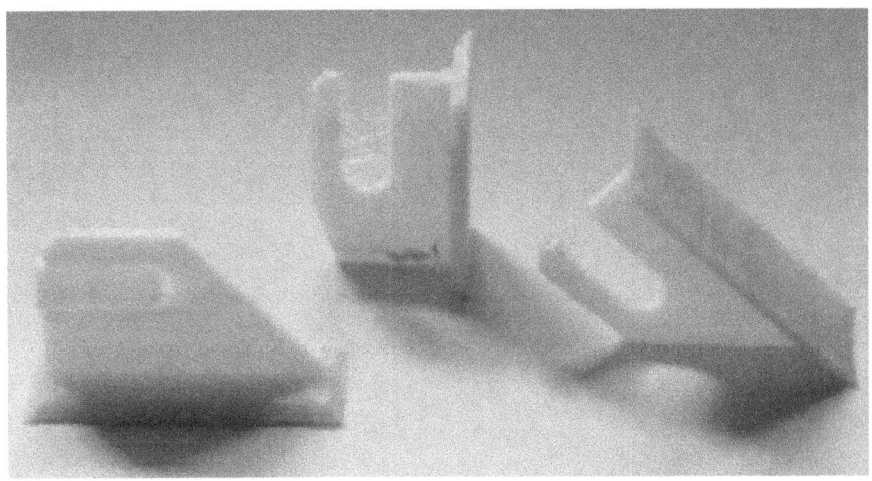

The support just pops out.

This is not always case on some prints you will need pliers to rip the support off and a file or abrasive paper to clean the faces up.

WHAT NOW?

Probably the best thing to do is download some bench marks or test pieces, there are many but basically you want a temperature tower, and many be an all in one 3D Print test.

Temperature Tower is as it's name suggests a Tower, of which each section is printed at a different temperature. The lowest level will be the highest temperature(so that it sticks) and then it will drop 5ºC each level it goes up. The range of Temps will be dependant on the material you are printing with. PLA it will be 220-185, PLA Plus would be 230–190 and PETG 260-230 etc. You'll probably need to do this every time you buy a new brand or type of Filament. You may think this is a waste of time but in doing this you will see and feel what too little or too much heat can do to your print finish. The obvious one is when the strands don't stick together because they were too cold to fuse. When you download this you will have to change the temperature at the various layers in your Slicer yourself.

All in one Test will show up over extrusion, bad temperature stability, stringing, belt tension plus at what overhang angle your printer need supports.

There are test for checking overall size, obviously this is important if you want to create accurate parts. All you do is measure the test piece in X, Y and Z and if they come out correct great and if not you put a correction figure into your Slicer. Never be tempted to modify your 3D model to correct this or else only that printer will be able to print it correctly and only in that orientation.

NOZZLES AND LAYER HEIGHTS

Obviously at some point you may want to create something more defined or conversely you might want to make something large that doesn't need to be so detailed. In which case you may want to change the size of your **Nozzle** smaller down to 0.25 or bigger up to 0.6. You can go bigger or smaller than this but don't jump the gun try sizes in small steps unless someone can assure you it will work with your printer and filament. Some filaments have contaminants in them that will easily and frequently block a nozzle. No such problems with a bigger nozzle but you may find your hotend can't keep the increased amount of filament hot enough to get it out and bond it before cooling soon. Also with bigger nozzles make sure your extruder can deliver enough filament to the nozzle or you may have to slow down your print speed so it can keep up. If you double the size of your nozzle you will need to get 4 times the amount of filament through it to print at the same speed.

Generally it's said you can have a Maximum layer height of ¾ (75%) of your Nozzle size and a minimum ¼ (25%) so

Nozzle	Layer Height Max	Std	Min
0.1	0.07	0.05	0.03
0.2	0.15	0.1	0.05
0.3	0.23	0.15	0.08
0.4	0.3	0.2	0.1
0.5	0.37	0.25	0.13

0.6	0.45	0.3	0.15
0.7	0.52	0.35	0.18
0.8	0.6	0.4	0.2
0.9	0.67	0.45	0.23
1.0	0.75	0.5	0.25

Some say you can go up to 80%. Most people will use their printers set at 50% height level so, 0.2 for most should give good results. You may find you can go passed these limits but it's probably better to change the nozzle size

Smaller nozzle doesn't always mean better prints, depending on the resolution of your printer and of the .STL file a finer nozzle may show all the flaws that a bigger nozzle would have smoothed out. Also the smaller nozzle will show up any flaws in your printer so you may have to do some upgrades to improve the finish. Don't forget the flaws you see may also be in the CAD file. If you see things like straights around your curves it can be the settings in the CAD software need changing to smooth this out.

Remember you can use a small or big Nozzle with exactly the same level height, the bigger Nozzle will allow higher angles of overhang. Below is a typical section through a single layer at 45º.

Above both are 0.2 mm height at 45º but the 0.4 Nozzle on the left barely touches, between layers, whereas the 0.6 mm Nozzle on the right has an overlap of nearly 0.2 mm

The other way to improve printing overhangs is to reduce the layer height. Using the same example as above but both using a 0.4 Nozzle and this time comparing it with 0.1 mm layer. Again the 0.2 mm height barely touches but the 0.1 mm layer height has a 0.2 mm overlap. Although this has the advantage of sticking your print together it comes at the cost of Doubling the time taken to print the object due to double the number of layers to print.

From this you can see you could also increase the overhang angle as long as are not printing at too high a layer thickness.

Layer Heights part 2 – I only came across this the other day so can't say how much difference it would make to your prints. Check the number of steps per revolution on your Z axis stepper motor and the pitch of your Z axis screw. Mine is 200 steps per revolution and a 2 mm pitch with 4 threads so that works out at 8 mm per revolution or true Pitch.

Divide the true Pitch by the steps per rev. Mine works out at

8 / 200 = 0.04 which any layer I put in should be divisible by 0.04 so 0.08 mm or 0.12 mm not 0.1 mm are ok and 0.2 mm is fine but 0.3 mm again is not, I need to use 0.28 mm or 0.32 mm.

I know these are only small differences and to most people it will not matter but there will be a pattern as it rounds up or down to its nearest step. So just be aware that if is see a patterns of some layers looking like a barcode rather than evenly space this could be the cause.

Some filaments will require you to use a bigger Nozzle or they will block as they contain particles of the material in them, be it wood, metal, nylon, carbon or whatever. I have used wood in a 0.6 Nozzle with no problems but my 0.4 did block, hence the 0.6.

Your **Slicer** should work out the correct feed rate given your Nozzle size, layer and filament diameter, that's why you should always check the diameter of you filament especially if you're trying a new brand but really check every one you buy. I'm sure different batches can vary and sometimes even over the length of a roll it can vary.

Also check the amount your extruder pushes through is correct, there is a place in most **Slicers** that you can correct for this.

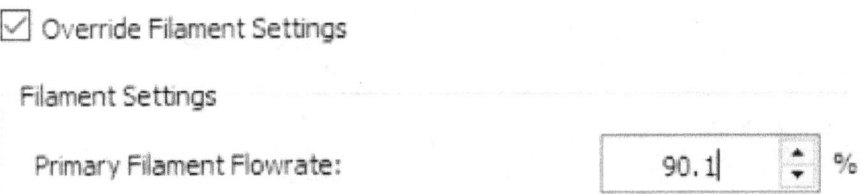

If you get 111 mm when you tell it 100 mm.

Take the amount it was supposed to feed and divided it by the amount it actually feeds and times by 100.

100/111 = 0.9009 x 100 = 90.09%

Round it up to the nearest 0.1 = 90.1%

Obviously if it only feeds 90 mm of filament you'd want to change it to 111.1%

100/90 = 0.1111111 x 100 = 111.11%

Rounded up to the nearest 0.1 = 111.1%

How do you change the Nozzle? Most printer are similar, turn on your printer and go into settings and turn up the Nozzle temp, above your normal print temp and let it heat up. Once it is up to temp switch it off, unplug it, and using 2 spanners hold the Heater block(aluminium block), don't snag the heater or Thermistor or their wires, and turn the Nozzle. When you fit the new one put it in finger tight then heat it up again and tighten it with the spanners. To tighten you only need to pinch it up, not full on weight behind the spanners. I use an inch tee so that I can't over tighten it. If it is too loose you will notice filament leaking around the Nozzle, if that happens tighten it just a little more. Only turn the Nozzle, the spanner on the heater block(aluminium block) is there to support it and stop you twisting it out of it's mountings. Some printers will require some disassembly to get enough access.

The photo above shows how it shouldn't be fitted, the following one is how it should look.

The Nozzle shouldn't end up locking into the block, there should be a slight gap between the head of the Nozzle and the Heater block. The Nozzle should be locking against the end of its thread face to the Heatbreak. So check the length of the threaded part of your new Nozzle against your old one if it is shorter you'll need to wind the Heater block further up the Heatbreak. This may require a bit more disassembly. As I'm using a Chinese Printer I ordered a few Chinese Nozzles and I have found them to be almost exactly the same size to the point that I can change from 0.4 mm to 0.6 mm and when I go to level the bed and set the height it's still correct. I still check, just incase.

Ok why are we worried about this? If you can lock the Nozzle into the Heater block and the is no gap then the Heater block is loose on the Heatbreak and effectively the Nozzle can move around (only a bit) whilst you're printing. This also means liquid filament can leak out the top and that the Nozzle can be difficult to get out of the Heat block due to the amount of metal to metal contact.

The gap doesn't need to be large, the general rule seems to be (if you are assembling the complete Hotend) take the Heater block, screw in the Nozzle and then undo it quarter to half a turn then fit that on the Heatbreak and tighten the Nozzle holding the Nozzle Heater block. And of course make sure there is still a gap..

I created the following part to test mine, it's 20 mm x 10 mm and 10 mm high. I made the wall thickness 0.4 (the nozzle size I was using) and put a circle and square in the walls. I had to make the base 0.8 mm as at 0.4 I ripped the bottom off. Later I found I was too impatient and if I'd just waited until the bed cooled a little more it would have released itself. My PETG one came out perfect but the PLA one was falling apart and the wall was too thick, so I reduced the filament flowrate to 75% I also increased the Nozzle temp to 217 from 210 and it came out in one piece with the correct wall width.

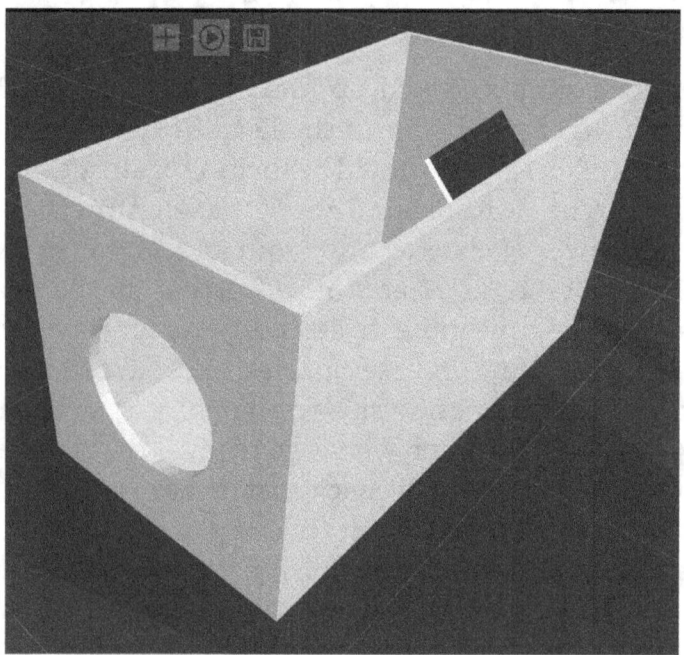

I was going to try the thickness test on the Slicer Cura but it would not allow me to print anything with a single wall thickness, it did the base and finished there.

I also created a large flat test piece, 300 by 300 by 1 which proved X was 0.01% too big and Y was 1.66% too small so I went into the Printer settings in my Slicer and put an X-axis compensation of 99.99% and Y-axis compensation of 101.66% an that seem to cure the inaccuracy.

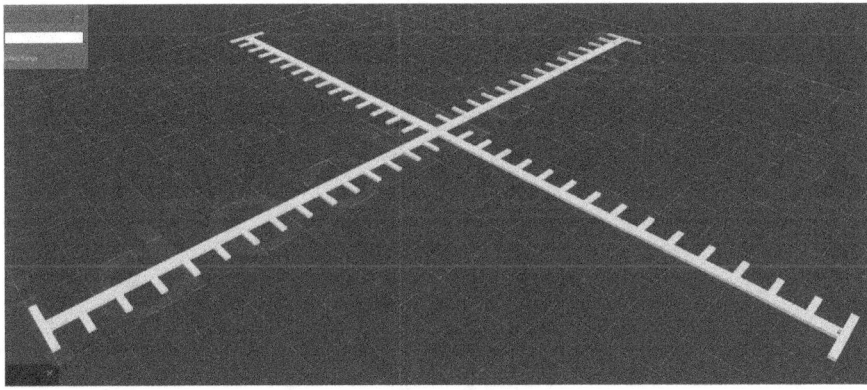

You may find that if you print the same part in a number of positions they come out slightly different size. My Geeetech A30 did that and after some thinking it became obvious. The X axis belt drops down 6 – 8 mm from the drive pulley to carriage pick-up , this means the belt may be at an angle of 15º near the motor and just over 1º at the other side of the printer. There was an easy fix, I created two blocks which raise the belt pick-up point so it's nearly straight between the two pulleys.

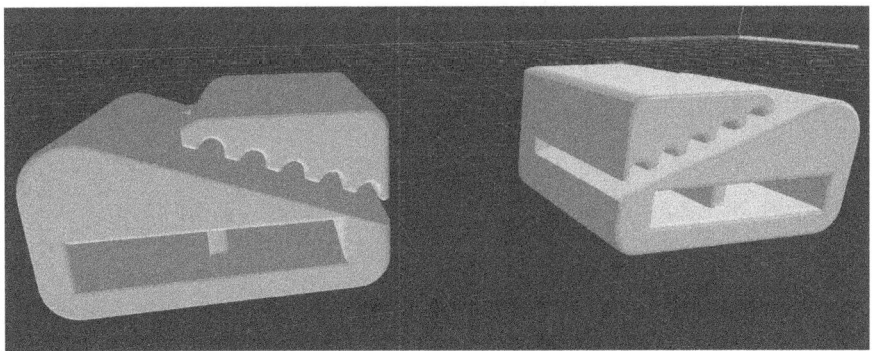

These slide onto the carriage where the belt originally fitted and the belt push through the toothed slot raising the belt height by 6 mm and locking in the belt.

More on this later.

WHAT NEXT?

Well this is where I started looking for parts to upgrade my Printer with parts I could make. I started by creating some sprung feet.

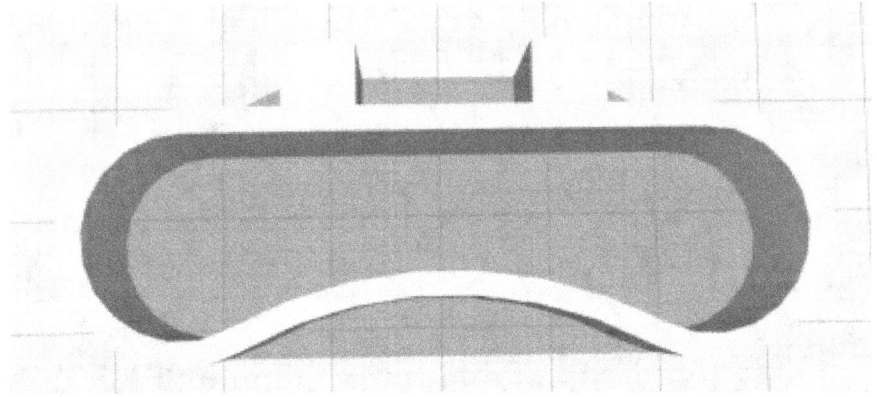

I started with 4 of these but still got noise, so I tried putting some foam under them. The printer was too heavy and crushed the foam so I printed 2 more. I ended up with 8 in total with 6 between the upright and the rear and only 2 at the front. It is a lot quieter and judging by the amount the foam squashes it now has an even loading on all the feet. Already I'm thinking of replacing these with some sockets that will take a squash/racket ball as I've seen online that they reduce noise even better.

Next I made some Bed Adjusting wheels

There were some on Thingiverse but they either didn't fit or were smaller than I wanted so I Designed and Printed my own. Remember to check what the thread is on the knob, a lot are M3 x 0.5 pitch so if you're putting notches around your adjuster you can have 10 notches and then you know as you move it one notch you've raised or lowered the bed 0.05 mm. Or 20 notches and moving it one notch has move it 0.025 mm. This means if you measure the brim or skirt of your prints it is easy to raise or lower a corner quite accurately.

Next was an Extruder feed wheel., so instead of pushing or pulling the filament through, I can wind it and also you can easily see that the extruder drive is working or not. I used the same basic shape that I created for the Bed Leveller and then just scale it down by 50% and recreated the centre section with a D shape hole to fit the shaft end.

Next a cooling cowl for the part. Now when you get your printer it will have a fan and maybe a duct blowing air towards the nozzle. Obviously you don't want to cool the nozzle only the part around it that you've laid down. Some have twin fans and ducts but I found a Cowl on Thingiverse that goes right around the nozzle but directs the cooling air down. Don't be fooled by the term nozzle cooler, it should be called part cooler as this is not to cool the nozzle but the filament as near the nozzle a possible, so it keep it's shape.

Next will require a screwdriver, a multi-meter and someone who knows about electric. If your motors get hot the first thing to check the Voltage output from your controller. It has been known that some of these are putting out nearly double the voltage required. These are easy to check and adjust but don't try unless you know what you are doing. The box will have mains power going in and it can burn or kill you if you touch the wrong parts.

Some people say you should get a Dry PTFE spray and treat the inside of the Bowden, Extruder/Hotend and Nozzle. This may reduce jams and increase the life of some parts. After swapping the 0.4 mm Nozzle for a 0.6 mm Nozzle I sprayed the inside of the tube, Hotend and Nozzle and still have not had any problems 12 months later.

I bought a "Micro SD to SD Card Extension Cable TF Memory Card Adapter Extender Converter", this plugs into the MicroSD card slot and leads to a SD Card receiver. This means I can plug it into the control box and then just use a larger SD Card. As it was only a matter of time before I pushed the Micro SD card in and missed the receiver. I then printed a holder (See below) for the SD Card receiver and stuck on the side of my controller, facing forward, with double sided sticky tape.

This may or not be of interest , I made a coffee table for my printer, this sits on my table/bench and my printer sits on it. What can I say I seemed to spend so much time bent over watch the first layer, or trying to look up at the nozzle for one or another reason I thought it the print was 12-18" higher it would be so much easier. And it is.

This won't affect everyone but one problem I notice with my Geeetech A30 was that gradually I was adjusting one side of the bed every few days until one side was 2 mm higher than the other then it stopped. The obvious thing to do was level the X axis beam/gantry with the frame and then re-level the bed. Doing this I notice that winding one Z lead screw to level the 2 sides it was getting tighter as I got nearer to level. I did it but again gradually I had to adjust one side on the bed until it was about 2 mm lower than the other and then it didn't seem to need so much levelling. I went on various forums and the main suggestion was to up the voltage to the stepper motor that appeared to be skipping steps. This didn't make sense, why would it do this and stop when it got to 2 mm difference? I didn't want to do it but the only answer I could see is that the beam had been assembled wrong at the factory and that one end was higher than the other. I started stripping apart the idler end of the X beam, if I was wrong when it came loose it would drop by 2 mm and be level. If I was right it would stay where it was. Sure enough it stayed 2 mm higher than the other side, which meant the other side had to be loosened off and adjusted until it was level and then retightened. Just one problem the vertical frame was between the screw heads and the holes in the Z carriage, that so say, gave you access to them. This meant I would have to take it off adjust it a bit, put it back on try check it and probably do this a number of times until I got close to level. Or I could put the level gauge under the X beam and drill through the holes and I'd be able to adjust them. That's what I went for. I set the levelling block under that side put a drill the same size as the holes through and marked the vertical frame and then took it off and drilled it through. I cleaned and reassembled it and then slacken off the screws and let the other side sit on it's levelling gauge. I tighten it up and the other side and reassembled all the bits I had removed and then levelled the bed.

That was over 6 months ago and since then it has worked great and only occasionally needed re-levelling. And both my Z leadscrews are still aligned, before they were 90 degrees out when it stopped needing adjusting. Another thing that came out of this was, when I first got the printer I'd tighten up all the screws that were loose but I had to slacken off the ones on the anti-backlash Z nuts as it didn't want move easily with them tight, Without thinking I tightened them up and it works fine.

The other thing I found on mine and I think it's probably true of the CR 10 as well, and that is the X axis drive belt comes down at an angle. I spotted this early on and made new belt ends that lifted it and got rid of a lot of the angle but not all. You might think what difference does make. Well first it means the further it travels away from the stepper motor the better the accuracy. And secondly because the belt angle changes the belt tension also varies. So I decided to create a part that fitted on the front of the stepper motor with a toothed idler pulley to drop it down below the X beam.

At the other end I fitted a tensioner I Printed from Thingiverse, https://www.thingiverse.com/thing:3530681. This was OK but the belt still didn't run parallel with the X beam so I redesigned the bearing carrier to drop the bearing down a couple mm or so now the belt runs as parallel as I can get it.

One other thing I found was that the Nozzle would not go the far side of the bed so moving it manually I notice it couldn't, the nut on the far carriage wheel stuck out too far and would hit the Z carriage. This was an easy fix, undo it and feed the screw in from the other side and put the nut on the front side of the carriage. Job done except I guess Geeetech knew this was a problem and limited the movement so it doesn't hit. I then had to download Python and Smartto Motor Tool to reset the limits which worked great.

The other thing I did was to create 2 packers, one fitted on the carriage where the X limit switch touches

And one a simple tube/cylinder that fitted on a bolt that the Y limit switch touches under the bed. These were made to stop the nozzle so that it's 1 mm in front and to the side of the bed when it homes. Then I told my Slicer that the bed was 2 mm bigger than it actually is and that home was X=0, Y=0 and Z=0. This works great the Slicer by default places every thing in the centre of the table and that is now where it prints it. If I'm only printing a small object I move it to over my front-left bed adjusting screw, if I do need to adjust the height I just adjust the one screw until the skirt is perfect. Obviously I re-level the bed before I print anything bigger, if I needed to adjust it.

LOCATION, LOCATION, LOCATION

Ok where are you going to put your Printer? For good printing it needs to be somewhere out of draughts, not damp, and not too cold. And before you say spare bedroom remember some of the filaments give off smelly fumes also unless you silence it or only do quick prints you may be able to hear it printing through the night.

Also for the sake of good prints it wants to be on something level and solid. This will not only cut the noise down but also stop vibrations coming back through whatever it's on and shaking the printer. And affecting your prints.

UPGRADES

It would be quicker to say what you can't Upgrade. Basically there will be a better or different version of everything on your printer that you can buy. Before you buy anything check it will fit your printer and only buy if it has a lot of good feed back.

SELF LEVELLING SENSOR

It's not something I want, I think it's something else that can go wrong. But if you struggle to level your bed it might be worth the investment.

You should still level your bed manually first, what the sensor does is maps the waviness of the bed. So as it prints it will move up and down with the bed to get a constant thickness. It will continue doing this, with a little less height variation each layer until it can print a whole layer with no change in height. It does not level your bed in anyway, it only maps the surface and allows the nozzle to initially follow that. So if your bed were on an angle your top printed face will be at an angle to the bottom face and not parallel. Hence the need to level your bed manually first.

NOISE

The first things for most people is the fan noise, it's the things that everyone seems to complain about. These are readily available and cheap. Next and also under the noise heading is Stepper Rubber Vibration Dampers, most people only buy 2 one for the each of the X and Y motors and leave the Z motor/s and Extruder.

Print some sound reducing feet, some people use squash/racket balls. Or foam pads under the feet.

REMOTE CARD READER

Micro SD to SD Card Extension Cable TF Memory Card Adapter Extender Converter, this does away with the Micro SD card and means you can position your new SD Card receiver where you find it more accessible. As you no longer have to worry you'll miss the Mirco SD receiver slot and either lose your card in the control box or damage either of them.

BEDS

You can buy a heated bed for ones without one or ones that get hotter or ones that heat up faster than your original one. Also you can upgrade the power module that powers it.

There are also a number of different tops to the Beds Glass, Flexible, Magnetic to name but a few.

My A30 Geeetech one is made of Silicon Carbide Glass with a micro-porous coating, it's great everything seems to stick when hot and when it cools down they fall off.

NOZZLES

You can get hardened, Plated and even ruby nozzles and don't forget different sizes, you don't have to stick with the one that was in it when you bought it. If you print mainly small items it might be worth getting a smaller nozzle and vice versa if you want to do larger items. Some people use bigger nozzles for everything and say the slight loss in resolution is worth the speed in which it prints. So you have to try and see want suits you.

I guess whilst were talking about the nozzle, the Extruder and Heater Blocks can also be replaced. And if you change the Heater Block you may want to change the Thermistor, Temp Sensors. Oh and maybe the heater block cover. And not forgetting the cooling fans on the heater Block.

One of the things you may hear is that your nozzle is going to wear out so you need a better one. Everything will wear out but not any time soon unless you are printing with something very abrasive like with Carbon fibre or metal particles and even then it will depend on who's Filament you are using, not all are as abrasive as others.

You can buy a Dry PTFE Spray that will coat the inside of your nozzle and heater block, they are not cheap but should reduce blockages.

I guess it's worth mentioning you can get Multi input Hot ends so you can change colour, I don't want this at the moment and think it's probably just something else to go wrong. Plus you have to set up for 2 different materials to be printed.

Don't forget the easiest mod is still to get different size Nozzles, they are not expensive but can make your prints finer quality with smaller ones or faster if your printing something large that doesn't the same amount of definition.

A while ago I saw someone using a 2 mm Nozzle on YouTube, yes 2 mm not 0.2 mm! He did some good single shell vases. He said his only problem was getting a good first layer but after that it flew up and was strong. He also said it ate filament, as it would do.

It's worth buying a cheap set of nozzles from china with cleaning needles, you can get a range (0.2, 0.25, 0.3, 0.4, 0.5, 0.6, 0.8, 0.9 or 1 mm) for a couple of Dollars, I've ended up using 0.3, 0,6 and 0.9 mm and have seen bough a some extra packs of just those as spares. Remember anything that has been designed by anyone else will be design to use with there nozzle so if you use thingiverse you may want to swap back to a 0.4 mm nozzle that's what most are designed for, On large objects this will not mater but if you have a 0,6 mm nozzle in you can't print anything with a vertical 0.4 mm wall thickness.

Z MOTORS

Strangely some people want to get rid of one Z motor and cross link it with a belt, whereas others want get rid of the belt drive and fit a second motor? You can change any of the motors for higher torque or smaller step i.e. 400 steps per rev instead of 200 which will give better accuracy, or both.

BOWDEN TUBE

As with the Z motors, some people who have Bowden Tube feeders want to fit direct feeders and some with direct feed want Bowden tube feeders. There are plus' and minus' on both sides. But unless you know someone with the same printer who prints similar objects to you then it's a gamble. The Bowden tube and the end fittings is something lots of users change. The standard Bowden tube can burn at the higher temps you need for some materials, but there are other tubes, Capricorn, which reduces friction and can withstand higher temps. It's a lot dearer but a lot of users say it works.

These are the most likely things you might want to change but everything can be, bearings, belts pulley wheels and all the electronics, you can bet someone has an upgraded version for you to buy. I would say stick with what you have until it becomes a problem and then upgrade.

On the upgrade note the old adage "if it ain't broke, don't fix it" springs to mind but the one thing I haven't done with my A30 is updated to the latest firmware. Reading online lots of people did and their screens didn't work after, so they were then told they had to dismantle the control box break out a micro SD that was glued into the back of the screen and burn a new version on to that and put it all back together. Mine works I won't be getting the latest version until I absolutely have to. I'm more likely to buy a new 32bit controller board, which has built in Super quiet TMC2660 stepper drivers, up to 256 microstepping. Takes a full size SD card, as well as loads of other advantages.

I'm not advertising but https://e3d-online.com are a great source of quality components and lots of documentation so you can thoroughly check what you're buying first. Also they a great source when it comes to creating your own parts to work with theirs as fully dimensioned drawings of most the items they sale are freely available on their site.

COOLING FANS

These are one of the most complained about things, noisy cooling fans. There are loads of quieter fans available to replace the existing ones. Just check the size and voltage, sizes vary and the voltage can be 12 or 24 volts. .

One mod I heard about but haven't tried yet is wiring in a 45º thermal switch it the control box so its cooling fan doesn't run constantly when switched on.

CABLE CHAINS

There are loads of these on Thingiverse and other places on the web and probably a few for your printer. Do you need them depending on the movements of your printer and your cable runs possibly not, but most of us will want them. They fix to the chassis of the printer and a moving part and you then clip your cables in them. This protects them from rubbing against anything and hopeful spreads the movement through the length rather than one part of the cable being moved/bent backward and forwards until the cables start to fracture.

BED ADJUSTER WHEELS/ EXTRUDER KNOB

You can buy, download or create your own adjuster wheels to make manual bed adjustment a bit easier and more accurate. And the same with a manual Extruder Knob, this has 2 uses, one to manually feed filament to or through the Nozzle to prime it or whatever. And 2 you can see that it is feeding or retracting or doing nothing, at a glance.

BELT ENHANCEMENTS

Obviously there are better grades of belts, different materials etc. that are stronger, stretch less or last longer but also it's worth looking at the route the belt takes from the stepper motor to carriage. On my A30 it dropped down 6-8 mm so the accuracy changed. The further it moved away from the motor the less the angle and the more accurate it got. Also does your idler pulley have teeth or is it plain? If the belt teeth are pulled tight against a plain pulley it will do the belt no good replace it with a toothed idler pulley.

I addressed both these problems, as you can see I made a small frame to fit on the stepper motor bolts that hold a toothed bearing to bring the belt down below the X beam.

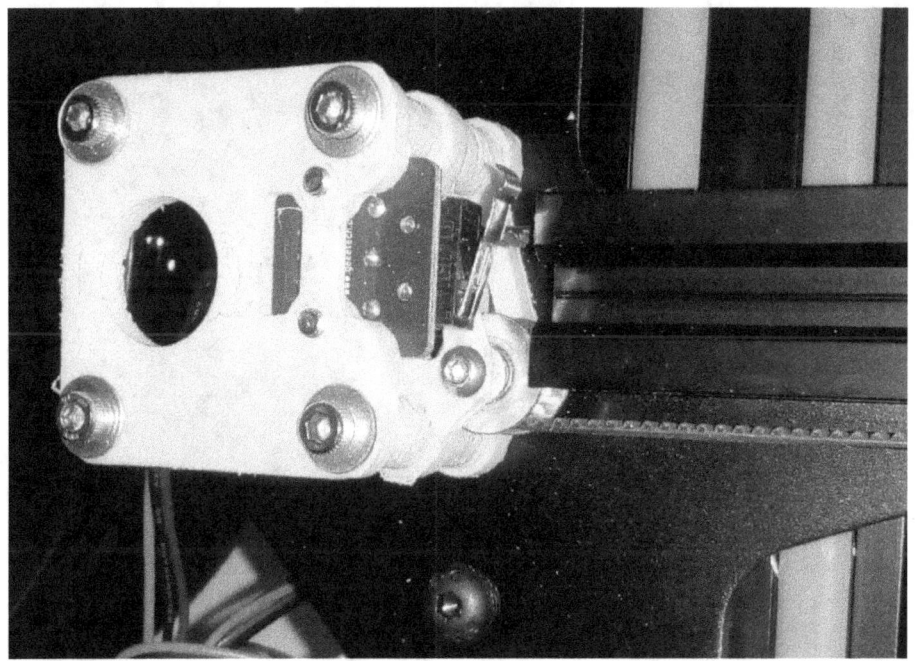

Then I printed a tensioner with another toothed idler dropped to the same level. Replacing the 2 piece smooth idler with another toothed pulley.

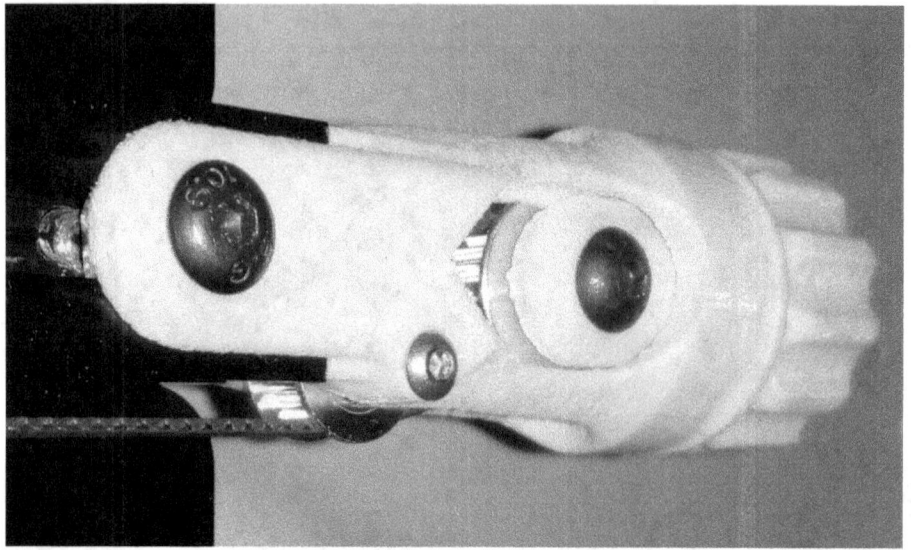

Next I printed 2 belt clamps that also raise the belt from the bottom of the carriage clamp to the same level as the 2 end idler pulleys. Now the belt runs parallel with the beam which means if the belt moves 10 mm around the stepper pulley where ever the carriage is it will now also move 10 mm.

WHAT TO LOOK OUT FOR

Elephants foot – not yours, if you print something that should have vertical sides but they bulge at the bottom this is probably elephants foot. The Bed maybe too hot and the weight of your object is squeezing it out at the base.

Pillowing – this is little sagging pockets you may get on a top surface. There are a number of reasons for this. First your infill is to little so that your top layers sag between them. Or secondly your top face is to thin. Either increase the infill or increase the number of top layers or a bit of both.

Stringing – this where you have fine strands between pieces. Its actually where the nozzle leaves one position and moves to another but due to the heat the filament still flows slightly. You might at first think this because your nozzle is to hot, that might have a part in it but usually its because you haven't got retract set in your **Slicer**. What is retract it a command that reverses the filament drive(extruder) just a little to pull back any molten filament into the nozzle.

Z Scar or **Zipper** – This is often where the printer starts and stops the layer each time. There are a number of options to get rid of this, in your **Slicer** settings select random or put a near to co-ordinate in. Find a corner and then it will be easy to sand out after. Apparently **Cura** has a setting called "**Spiralize outer Contour**" which is supposed to make the outside layer continuous all the way up your object. I don't know how this can be but if you use **Cura**, try it, it might work.

DANGERS

I know it's a Printer how dangerous can it be. To any children it's going to be magical and very Dangerous. The Bed can be anything up to 120ºC and the Nozzle 310ºC. Not to mention little fingers, sprockets and toothed belts.

OK that's the little children but what about us big kids yes we too can get burnt, once you're in Printing mode think about where you are putting your hands, it's easy to see some debris and want to brush it a side. Stop and think before you do. Also this is a mains electric machine so go careful if you mess with it when you have it plugged in and switched on.

Classic one is brushing up the bits and getting the brim, raft, skirt or supports off. Plastic can cut or cause splinters so beware. Also if you have to use a scraper or blade to get you object off of the bed mind your hands.

The most common injury is said to be stabbing yourself with the tool as you try to get the print off the bed with, so be careful or get a better bed

Last one is keep the Printer away from anything flammable, it might be worth fitting a smoke alarm where you have your printer. It is very unlikely that your printer would start or cause a fire but there have been a few case where it might have.

What has come light of late is the importance of making sure you have a good Earth/Ground, especially if you have a metal framed Printer. Check your socket and extension lead if you are using one, for a good earth/ground and make sure if you use your own power lead that it is a 3 wire one and not a 2 wire.

Why?

Because some of the power supplies used, leak some Voltage back to earth/ground, it may be only a low voltage and might not kill you but why take the chance.

This is not a reason to boycott all printers, this can occur on any electrical device plugged into the mains. They are usually either double insulated or come with a earthed/grounded 3 pin plug and that only works if it's correctly connected to an earth/ground.

WHAT NEXT?

I'd been using the PLA Wood filament a lot, still with the 0.4 mm nozzle and I was printing a 300 mm long ¼ scale pedestal(plant stand) as a demo for one that was going to be made. It was a 3 hr print so I left it and occasionally looked in. At about an hour I looked and it was looking good, I looked again an hour later and all seemed to be OK except it hadn't got any higher? The printer appeared to be working, the feeder wheel I made was turning but there was a gap between the nozzle and the print I stopped it and looked still everything appear OK , I wound out the filament and there was what I though was the problem the filament ended where it entered the feed wheel/cog, hence the detector didn't stop the printer as there was still filament in it. I tried pushing the filament in the Bowden tube but it wouldn't move so I had to remove the Bowden tube and pull the filament out. I had the nozzle heated but it was still difficult to pull or push. I did get it out but I couldn't clear the nozzle. I tried it with the heat up high and pushing it through and then leave it in hot and let it cool and pull it out. Nothing, it was blocked, at this point I thought I'd change over the nozzle to a 0.6 mm one as I had bought various sized spares. I raised the end to near the top and heated up the nozzle to it's max 250ºC and with a pair of water pump pliers holding the hot end/ heater block and used a 7 mm socket to undo the nozzle. It came out easily and so I removed the Bowden tube totally and spotted my blockage, there was about ½" (13 mm) of the tube from inside the Hotend that was black. I tried pushing a metal rod through but it was not budging. I put Acetate down the tube and it eventually started moving. Even after getting the cooked filament out new filament would not pass easily through, so I covered the wood filament in Acetate and kept pushing that through until it looked clean and didn't stick. I replaced the nozzle (now 0.6 mm), as I took it out, with the hotend heated fully. After it was nearly all back together still had the Bowden tube off at the feeder end, I sprayed Dry PTFE down the tube until it dripped from the nozzle. Don't let this get on your table or you may have problems. I left it with the nozzle set to 80ºC for 10 mins and then reconnect the Bowden tube to the feeder.

Ideally if you do this leave it for 24 hr before using it or you might get the first filament not sticking to the bed.

When I used it with the 0.6 mm nozzle it was a lot quicker at printing the same things. Don't forget you must change your **Slicer** settings when you change the nozzle size, and most important re-level the bed, mine was still level but I guess 0.2 mm further away so the whole bed needed raising.

I had to run my test pieces again for each filament now I was using the 0.6 mm Nozzle and found I had to reduce the Filament Feederate to 75% to get the wall thickness correct. I also revisited my table levelling method. The first layer always looks too thin. So I peeled the skirt off and checked it's thickness with a vernier, sure enough the first layer which should have been 0.2 mm was coming out 0.1 mm so I set the nozzle height on the Printer to 0.1 mm and checked the bed level using my 0.2 mm piece of paper and on the next print the skirt did indeed come out at 0.2 mm? This also explained why my heights were coming out slightly short.

You will find you'll be doing a lot of the same things over and over again to get your printer to give good results.

A few things to be aware of, don't move the table about by hand whilst it switched on unless you disable the stepper motors. As with most motors they will act as a generator when moved by hand and so you can damage the circuits if you quickly move them.

If you're thinking of designing your own cooling nozzle make sure you design it to cool around the nozzle not at the nozzle itself. Recently I've seen a few downloadable cooling ducts that seem to point at the nozzle. The idea is to cool the print not the nozzle but the filament the moment it leaves the nozzle. If you point it at the nozzle you may find you need to increase the nozzle temp to get your prints to stick together.

A good test is to heat up your nozzle and feed some filament through, it should go straight down. If it instantly curls up then you have a partial blockage. If it instantly veers away from your cooling duct (if it's on) it's pointing in the wrong direction. If a couple of mm down it veers away from the outlet that's should be fine. You can also check the diameter it is coming out the nozzle at and don't forget to push through the new Filament when you change from one spool to another or you may get some lumps, voids or colour bleeds when you start your next print. If you have 3 or 4 skirts around your print then this will do a similar job.

Always feed the Filament in right the way to the hotend and wind some through as the nozzle gets up to temp before starting your print.

I don't know if it because the weather is getting colder, I know it shouldn't make any difference but I had 3 prints move whilst printing I couldn't find anything wrong anywhere but I thoroughly cleaned the bed again and check it's level. I ended up by increasing the bed temp by 2ºC and no more problems. Why this happened I can't say other than the weather had turned cold so I thought it might be worth a try.

Wow it happen yet again none of my prints would stick, even ones that had the day before. So I cleaned the bed thoroughly still wouldn't stick. Started levelling the bed and there it was, either the nozzle had some how moved up or the bed had moved down? I expect it after changing something, or it gradually changing but to suddenly change over night I don't understand. Clearly levelling will be more part of my life than I originally thought. Anyway reset the level 0.1 mm on the printer and using a 0.2 mm feeler gauge and all was well again until the next time. Don't forget to make sure your Nozzle is clean when you check the level a second or third time. The Nozzle may have cooled filament on it and give a false height, if in doubt heat the Nozzle up and clean off any debris.

WHAT TOOLS DO I NEED?

A lot of printers come with some basic tools like small screw drivers and Hex(Allen) keys, which are ok for checking things are tight but you will need more at some point. If you want to change Nozzles you'll need a socket and possibly more than one if the new one is a different size my original one was 7 mm and the new one was 6 mm others maybe larger. You'll also need some water pump pliers (not normal pliers) these will grip your hotend between parallel jaws and not on the edges as normal pliers ones would.

Some snips/wire cutters or even a pair of scissors to cut the filament off with.

Depending on your Printer Bed you may need a scraper to get the print off.

You'll want some cleaning materials, Acetate, Isopropyl or Alcohol swabs or liquid and lint free cloths to keep your bed clean.

You need some type of measuring equipment, a Vernier is probably the cheapest and most versatile thing you could have, as you will need something to check layer thickness and filament widths. You may say you're an artist I don't need accuracy but even if you are creating art you probably want a circle to come out circular rather than elliptical. And if you want to replace an existing item or make a new part to fit something that already exists you will need to measure these parts accurately.

You'll want some kind of deburring tool to clean off brims and clean up any sharp edges or bits.

Some people use infrared temperature gauges to check bed and Nozzle temps.

Some people use hot/heat guns to melt off any stringing that occurs.

Wet and Dry abrasive paper is worth having to clean up Prints. Careful if you use a power tool you may melt the print, so sand it wet and if drilling, drill slowly and only for a short time and cool it with water and then drill a little more and don't use high speeds.

MISCELLANEOUS

This is just a few things I've found useful to me, you may or may not find them equally useful, I hope you do too.

If you are trying to print an assembly with moveable joints print a test joint first, you may need to increase the clearances(gaps) between the parts and remember to check your Slicer preview to make sure there are gaps. If your layer print thickness is 0.2 mm don't expect a 0.1 mm gap to work. You will find a clearance that works in one material will probably not work in all materials. You will need to break the joints when you first use them, if they are not a little bit bound together then you may have too much clearance for some joints to work correctly. It is usually easier to create separate parts and assemble them later unless they can only be assembled by printing them together. And don't forget to work out your optimum Z steps and make sure your gaps start and stop on a layer joint not in between or you may have no gap.

Naming your G-code files, you could number them 0001 and so on, and keep a list of what they are, but then if you want to print one again later you have to look it up. I use a description but be aware some characters won't show on your printer display. Also are you going to be using different filaments or nozzle sizes? Then you either need to get that in the title or keep them in separate folders.

I use "PLA 06 suzuki wheel mk3 Pair centre cap", "PetG 04 test 10x20x0_4" or "Wood 04 test 10x20x0_4". So you can see I start with the Material then the Nozzle size and then file name. I did start to include layer height and infill percentage but on my display you only see the start of the name so it seemed pointless.

One thing on my A30 printer, I can have loads of gcodes on the microSD card but it only displays the last 28 files. There are 4 files shown at a time and it only pages back 7 screens. Other printers may work differently but just remember the file you want to print maybe on the card but just not visible. The same goes with copying across an old file to print again if the date code is older than the 28th oldest one I won't see it on my printer.

Obviously if you stick with one Filament and one Nozzle you need not worry and if in the future you do change then you could have all PLA 0.4 Gcodes with no prefix and just use Prefixes for anything different. As I said you could just put them in different folders, the problem with that is when they are on the Printer you won't know what they were written for. You just need a system that will work for you. Remember that a couple of years or month even you maybe using a different brand of the same Material that may well require a different Gcode.

Speaking of Slicers I updated the Slicer software and thought nothing of it all my templates were still there so all was well. I'd decided to Design and Print. A modified moving, hinged, assemble. First try came out solid so I decided I'd increase the clearances/gap between the mating parts, printed it a second time, solid again. I increased the clearances to 0.2 mm all around and still solid. I looked at the part and noticed some groves or gouges in a few places, so I ran it again and paused it after a few layers. At this point it should be 3 separate parts, a pin surrounded by one part and then the second part that would join the pin after a while. But they were joined. I watched the levels print and they were separate but then the Nozzle, no filament that I could see, some how gouges across the parts. I went back to my PC and looked at the Gcode with Notepad++, I did a **Find** "Z" and I could see the layer was increasing by 0.2 mm but then it would decrease by 0.2 mm, thus melt through and joining separate parts. My first thought was a bug in the new software so I uninstalled it and re-install the older version. I didn't print I just looked at the Gcode and it was the same. I remembered that I still had a copy of it on my old XP PC so I started that up called up the same file and sliced it. No change in the Z or at least it only increased no up and down. I looked through both and spotted that in my original setting I'd never used a setting called "Z Hop at Retraction" which lifts the nozzle when it passes over printed stuff. It doesn't it drops it down, any figure I put in drops it and it won't let me put in a negative number? I went back to my Windows 10 and checked the "Z Hop at Retraction" value it was 0.2 mm. I set it to zero and Sliced my assembly again, it came out fine, it a bit scary as you have to force it until it make a snapping sound and it moves.

After it worked I decided to update the software again and sure enough in all my templates that had just said "0" in "Z Hop at Retraction" it now said "0.2" I changed them back. I looked at the Gcode on the Slicer's Previewer and saw nothing but I switched on the show retract position and that was the places that the print was being gouged.

Further to the Hydroscopic Filaments thoughts, I use a lot of PETG and I was printing some cable clips which I had done on a number of occasions and heard the dreaded popping sound whilst printing. The parts didn't look great but when I cleaned them up and tried to snap a cable in, they broke. So I think I can confirm it is a problem with PETG so far wood and standard PLA have been fine.

Online someone came up with a method of drying out filament and that was to put the filament on the bed and put a saucepan over it and heat the bed up and leave it for an hour or so.

This may or not work for you but I recently decided to use mainly 2 Nozzles 0.3 and 0.6. Why? This means if I design my parts for the 0.6 and I need more accuracy I can print the same things with the 0.3 and at worse it takes twice as long. I could have chosen 0.4 and 0.8 but I felt they were bigger than I wanted. Ideally I'd have gone for 0.25 and 0.5 bit I think the 0.25 might block more. As it is if I find too many blocks with the 0.3 I may change to 0.4 and 0.8 as my main Nozzles.

When you're designing your Prints check the Slicer preview.

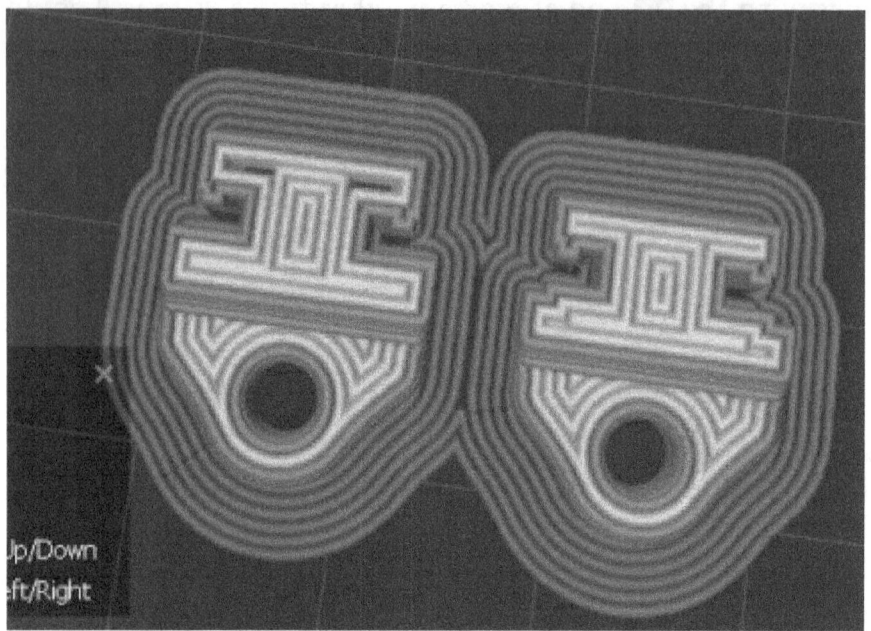

You may notice the print on the left has some voids in it so by reducing the size of the one on the left it is now solid, I could have increased it. The basics of this is to design for the Nozzle size you're going to use. So if you are using a 0.6 mm Nozzle then make you're your widths in your plan view(parallel to the bed) a multiple of 0.6 mm. It's not the end of the world but on small parts it will be a bit stronger without the voids. Don't forget you can rotate the items in the Slicer to get the layers going the way you want, even if you designed with a different orientation.

Same thing applies with the layers if you have a 1st layer of 0.2 mm and then continue with 0.2 mm then try to arrange your features on 0.2 increments if you require something to a specific size.

You may have noticed I'm using brims on these items, I did try without but on such a small item 12 mm x 16 mm they came unstuck, but with the added brim it was no problem.

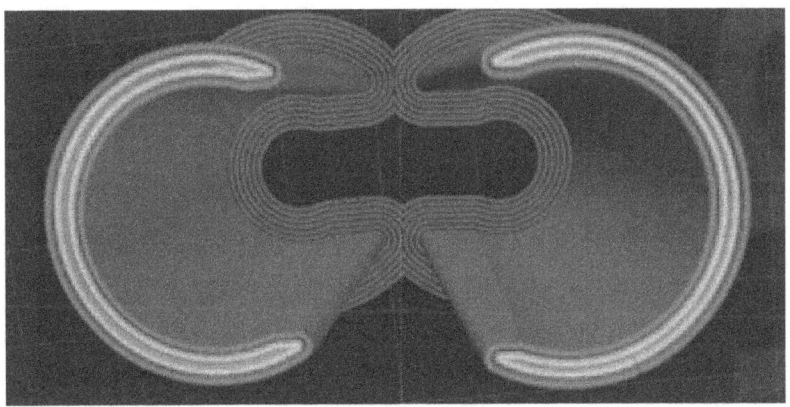

Another void avoidance, I was creating a conical piece to hold our shower head and in the Slicer preview I noticed a gap between the 2 inner shells. I could have changed to a smaller Nozzle but I didn't want to do that. I couldn't change the inner outer diameter as it would then be the wrong size. So I extended the part length by 1.3 mm which was the point where the inner and outer surfaces were 2.4 mm thus 4 times the 0.6 mm Nozzle widths and the end was solid.

So what do I use my Printer for, well just in the last month or so I've printer flower pot feet, twist in greenhouse cable attachments, level pot holder/ dish to go on sloping window sills, hose end holster to fit the green house. You have noticed a bit of a pattern there, my partner is into gardening.

A quarter scale 3D wooden plant stand for a local Carpenter to make a life size one from. A tapered fitting to allow a shower head to fit our shower head holder. A hair trap and riser to allow us to use the shower base as a foot bath whilst we shower. Various hair traps for baths.

Recently the Garden Club needed some Trophy stands as they were running out of space on the name plates, to buy one new base was £70 plus the cost of the nickel ring. So I printed some out at only a fraction of the cost.

Wheel centre caps, car badges, ignition cable tidies, connector for charging our strimmer/ brush cutter, new improved impellers for our pond, cable clips, cable covers, net curtain hooks, wheelbarrow wheel bearings and spacers, funnels, bag Clips, nut and bolt container, bouquet holding pew end hooks, Oasis cutter for pew ends, sdcard reader holder, mail box seals, mail box elevated draining floor, various knobs, pipe joiners, picture frame boarder, book ends, strimmer/brush cutter bump head, centre finder, fan duct, loads of test pieces and loads of odd bits. I think it's paid for itself.

Threads – generally I wouldn't advise printing threads I usually either cut the threads with a tap after printing or design in a captive nut or bolt. The other alternative is to use thread inserts and push/melt them in with a soldering iron.

Sometimes if it's a larger diameter as in a lot of the belt tensioners then both the nut and bolt parts can be printed. These are more often than not printed on their side, thread axis across the bed. But this does not have to be the case, there is a thread called Buttress which is not a normal thread, one side is cut at 45º and the other 0-5º which as long as it's printed up the right way is printable without any supports.

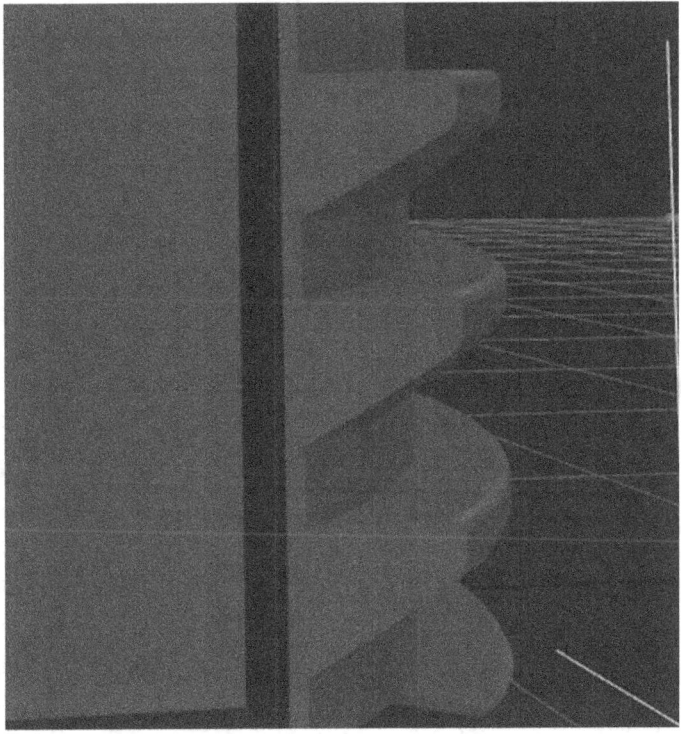

You should be able to print an overhang of 45º with no trouble but 50º will probably be your limit. After that you'll probably need to use support or change it's orientation.

SOME USEFUL LINKS

These are a few YouTube channels I've found useful and that had good advise, there are others that I didn't agree with and probably loads more that I haven't even found yet. These all cover loads of Printers but obviously you may want to find some that just concentrates on yours.

3D Printing Nerd - https://www.youtube.com/channel/UC_7aK9PpYTqt08ERh1MewlQ

Maker's Muse - https://www.youtube.com/user/TheMakersMuse

Thomas Salander - https://www.youtube.com/user/ThomasSanladerer

Jimmy Shaw's Tidbits - https://www.youtube.com/channel/UC8AnKTauPDig93avoC-zXSQ

WHAT ABOUT THE FUTURE?

Well in the short term, there will be more innovation with Filaments, that's what got us to where we are now and I'm sure it will keep happening. I would think the big thing will be recycling, at present recyclers will not take your cast offs as they don't know what it is, but maybe filament Companies will take your cast offs and offer you a discount on your next roll of Filament. Also home recycling will be better, at the moment there are a couple of machines that will chomp up your old prints and waste and melt it down and produce new filament. Of what I've seen they are not quite there yet with ease of use and reliability but then neither were 3D printers a few years ago we know it can change just like that.

Only this week I was reading that a company was 3D printing carbon composite conrods that are good for 15,000 rpm and 3,000 horse power! And they are 10 time lighter than steel and 6.5 times lighter than Al-Alloy! Not a home printer but give it a few year and who knows.

I'm sure there will be more innovations in 3D printers themselves, we can already see an automated swap system for print heads, pen holders, laser/tool holders, etc. so your 3D printer can do a number of things without you having to touch it. Prices will come down or accuracy and sizes will go increase.

There are conveyor belt type print beds, which rolls the finished item off and allows your printer to get on with the next print so you can get a production line going.

I would guess at some point the Slicer will be built in and self aware so it will test the filament as it is fed in and you'll just give it the file name and quality you require.

Mid term I would think we will be using lasers and powder or resin to create models at home although it might be while before this becomes affordable to the masses.

Long term no one could have guessed the rise of mobile phones/devices from the brick with a battery pack attached to what we have today. So whatever we think we're probably wrong, what will be is something remarkable, maybe wonderful and maybe not. Who knows we might be printing from rubbish or molecules in the air?

Buy one, get your foot on the ladder and don't miss out in the mean time.

G & M CODES

G-Code has been around for years (1950s or 60s) and is and was used to control CNC (Computer Numeric Controlled) machines, but has been slightly modified for 3D printers as they didn't account for hot ends, bed or filament extrusion. G is an old movement command and M was Miscellaneous functions or Auxiliary Commands.

Not all G codes work with all machines, most the companies run their version of it and seem to stay in some places with their commands. These are ones you are most likely to come across in your .Gcode files.

- X** The position to move to on the X axis
- Y** The position to move to on the Y axis
- Z** The position to move to on the Z axis
- E** The amount to extrude between the start point and end point
- F** The Feedrate per minute of the move between the start Point and end point, where Feedrate is actually the nozzle speed across the bed.

Code	Description
G0	Rapid Movement X Y Z
	G0 X10 Y10 Z20 – move to X10 Y10 Z20
G1	Coordinated Movement X Y Z E F
	G0 X10 Y10 Z20 E10 F800 – move to X10 Y10 Z20 extrude 10 mm of Filament during that time at a federate of 800 mm/s
G4	Dwell S(sec) or P(milliseconds)

G4 S10 pause 10 seconds

G28 Home each Axis
G28 X - go to origin of X
G28 X Y Z - go to origin of X, Y and Z

G90 Use Absolute Co-ordinates
All Co-ordinates are from the origin

G91 Use Relative Coordinates
Co-ordinates are relative to the last position

G92 Set current position to Coordinates given
G2 X0 Y30 – nothing moves but the extruder now thinks its at X0 Y30

M0 Unconditional stop
M0 - Printer finishes last move and then stops and shuts everything off

M18 Disable all stepper motors; same as M84

M84 Disable steppers until next move or set an inactivity timeout. Its now safe to manually move the print head or table

M104 Set extruder target temp
M104 S210 – set the extruder temp to 210°C but carry on, do not wait.

M106 Fan on
M106 S0 – M106 S255
Turn fan off (0) Full speed (255)

M107 Fan off

M109 Set extruder target temp and wait for it to be reached
M109 S210 - set the extruder temp to 210°C but wait until it reaches it before continuing.
M109 R220 - set the extruder temp to 220°C but wait until it reaches it before continuing also when it's cooling.

M112   Emergency stop

M140   Set bed target temp
            M140 S50 – set bed temp to 50ºC and continue.

M190   Set bed target temp and wait for it to be reached
            M190 S50 - set bed temp at 50ºC and wait until it reaches that temp before continuing.
            M190 R50 – set bed at 50ºC but and waits heating and cooling.

M220   set speed factor override percentage
            M220 S95 override the speed of the printer to be 95% of the defined speed

M221   set extrude factor override percentage
            M221 S95 override the extrusion rate of the printer to be 95% of the defined speed
            M221 S110 override the extrusion rate of the printer to be 110% of the defined speed

ABOUT THE AUTHOR

Nik was born in the late Fifties in Bristol (UK) and did an Apprenticeship as a Fitter/Turner 1975-79. On finishing that he joined Bristol Aircraft Corporation as a Trainee Draughtsman and Specialised in Lofting (Full Scale Layouts and Geometry). During the time when interest rates went through the roof (17% +) he went contracting (Freelance) to pay the mortgage. He worked on most of the Airbus range plus DeHavilland's Dash 8, Fokkers, HS 146, Hotal (space plane), Shorts, and much more. He spent years in the Wind-Tunnel Design Office and in "In Service Support Design" group.

When he was in his 40's he was told he was Dyslexic, which probably wasn't that much of a surprise to people who saw the reports what he wrote. In his 50's, after he and his partner watched a program about it, he took an Autism test and no one seemed surprised to find he was also Autistic.

In the '92 he started the Avon Sports And Classics Car Club and the Sprite/Midget help pages on the web (http://www.omniautos.f9.co.uk/sprite/index.htm) and HTML help (http://www.htmlcode.plus.com/index.htm) soon after.

Now in his spare time he runs a number of websites for various local organisations, has written a few books and collaborated on another, designed bits for rally/race cars as well as building them. And messes around with computers, which he has been doing since the 70's when you had to build your own and write your own programs.

He has given blood since the 1970's and is on his 340+ donation thanks largely to donating plasma weekly in the 90's.

He was trying to learn Java, to bring his programs into the 21st Century but was struggling due to Dyslexia, he was hoping to meet someone who could give some pointers after reading various books and getting nowhere. He finds he needs to use examples rip them apart, change bits and see what that does, to find out how they work. Not really the correct way but it works with most things. He has mastered the part he was after, so a user can input numbers, the program works out the answer and displays it. And he has put his examples into a book so others can benefit.

After being given an Android Tablet as a present and realising there was a lack of help around he wrote his popular "Android Tablet Basics" which has gained a good audience and is updated yearly.

A few years ago he discovered MIT's App Inventor 2 website and started using that to create Apps that calculate things and has a number of them on Amazon, both free and for cash. Again it was thought other users would be interested on how that's done so he wrote "Creating Calculating Android Apps, using MIT App Inventor 2".

His Tyre/Tire Markings book (ebook and now also in paperback) came about because he was trying to find out about the markings and although they were all out there somewhere, they didn't seem to exist in one place.

Nik started Designing for 3D printing back in the 90's and never dreamed he'd have one at home, back then they were around £750,000. This book again is an answer to the lack of 3D printer/printing info that comes with printers. And hopefully it will help answer the question, "Is 3D printing for you?"

As an ex Loftsman (Full Scale Layouts and Geometry Draughtman) Nik has been working on a book to explain how it's done and why there is still a use for it in this day and age. But it stops and starts between other things. It's one of those he wants to put out there but there is probably little interest, so it is a passion project and will probably take a back seat to all other projects.

www.ingramcontent.com/pod-product-compliance
Lightning Source LLC
Chambersburg PA
CBHW070421220526
45466CB00004B/1498